Editor
Lorin Klistoff, M.A.

Managing Editor
Karen Goldfluss, M.S. Ed.

Editor-in-Chief
Sharon Coan, M.S. Ed.

Cover Artist
Barb Lorseyedi

Art Coordinator
Kevin Barnes

Art Director
CJae Froshay

Imaging
Rosa C. See

Product Manager
Phil Garcia

Author

Mary Rosenberg

Publishers
Rachelle Cracchiolo, M.S. Ed.
Mary Dupuy Smith, M.S. Ed.

Teacher Created Materials, Inc.
6421 Industry Way
Westminster, CA 92683
www.teachercreated.com
ISBN-0-7439-3741-4

Reprinted, 2003
Made in U.S.A.

Table of Contents

Introduction . . . 3
Practice 1: Identifying Lightest to Heaviest . . . 4
Practice 2: Measuring in Inches (Length) . . . 5
Practice 3: Identifying Temperature . . . 6
Practice 4: Identifying Number Words . . . 7
Practice 5: Finding the Area . . . 8
Practice 6: Identifying Shape Families . . . 9
Practice 7: Identifying Shape Patterns . . . 10
Practice 8: Identifying Symmetry . . . 11
Practice 9: Identifying Position . . . 12
Practice 10: Adding Money . . . 13
Practice 11: Adding Money . . . 14
Practice 12: Counting Money . . . 15
Practice 13: Subtracting Money . . . 16
Practice 14: Adding and Subtracting Money . . . 17
Practice 15: Finding Equivalent Money Values . . . 18
Practice 16: Telling Time on the Hour . . . 19
Practice 17: Estimating Time . . . 20
Practice 18: Telling Time to the Half Hour . . . 21
Practice 19: Comparing . . . 22
Practice 20: Using Greater Than, Less Than, and Equal To . . . 23
Practice 21: Ordering Numbers . . . 24
Practice 22: Adding Using Tables . . . 25
Practice 23: Counting On . . . 26
Practice 24: Solving Math Riddles . . . 27
Practice 25: Subtracting Using Tables . . . 28
Practice 26: Identifying Sums and Differences . . . 29
Practice 27: Estimating . . . 30
Practice 28: Adding and Subtracting Numbers . . . 31
Practice 29: Using Ordinal Numbers . . . 32
Practice 30: Using Ordinal Numbers . . . 33
Practice 31: Using Numbers to 50 . . . 34
Practice 32: Using Numbers to 100 . . . 35
Practice 33: Solving Number Riddles . . . 36
Practice 34: Identifying Tens and Ones . . . 37
Practice 35: Identifying Tens and Ones . . . 38
Practice 36: Solving Word Problems . . . 39
Test Practice 1 . . . 40
Test Practice 2 . . . 41
Test Practice 3 . . . 42
Test Practice 4 . . . 43
Test Practice 5 . . . 44
Test Practice 6 . . . 45
Answer Sheet . . . 46
Answer Key . . . 47

Introduction

The old adage "practice makes perfect" can really hold true for your child and his or her education. The more practice and exposure your child has with concepts being taught in school, the more success he or she is likely to find. For many parents, knowing how to help your children can be frustrating because the resources may not be readily available. As a parent it is also difficult to know where to focus your efforts so that the extra practice your child receives at home supports what he or she is learning in school.

This book has been designed to help parents and teachers reinforce basic skills with children. *Practice Makes Perfect* reviews basic math skills for children in grade 1. The focus is a review of math skills. While it would be impossible to include all concepts taught in grade 1 in this book, the following basic objectives are reinforced through practice exercises. These objectives support math standards established on a district, state, or national level. (Refer to the Table of Contents for specific objectives of each practice page.)

- measuring in inches
- identifying temperature
- identifying shapes
- using money
- telling time
- adding and subtracting
- using ordinal numbers
- using numbers 1–100
- solving word problems
- identifying tens and ones

There are 36 practice pages. (*Note:* Have children show all work where computation is necessary to solve a problem. For multiple choice responses on practice pages, children can fill in the letter choice or circle the answer.) Following the practice pages are six test practices. These provide children with multiple-choice test items to help prepare them for standardized tests administered in schools. As your child completes each test, he or she can fill in the correct bubbles on the optional answer sheet provided on page 46. To correct the test pages and the practice pages in this book, use the answer key provided on pages 47 and 48.

How to Make the Most of This Book

Here are some useful ideas for optimizing the practice pages in this book:

- Set aside a specific place in your home to work on the practice pages. Keep it neat and tidy with materials on hand.
- Set up a certain time of day to work on the practice pages. This will establish consistency. Look for times in your day or week that are less hectic and more conducive to practicing skills.
- Keep all practice sessions with your child positive and constructive. If the mood becomes tense or you and your child are frustrated, set the book aside and look for another time to practice with your child.
- Help with instructions if necessary. If your child is having difficulty understanding what to do or how to get started, work through the first problem with him or her.
- Review the work your child has done. This serves as reinforcement and provides further practice.
- Allow your child to use whatever writing instruments he or she prefers. For example, colored pencils can add variety and pleasure to drill work.
- Pay attention to the areas in which your child has the most difficulty. Provide extra guidance and exercises in those areas. Allowing children to use drawings and manipulatives, such as coins, tiles, game markers, or flash cards, can help them grasp difficult concepts more easily.
- Look for ways to make real-life applications to the skills being reinforced.

Practice 1

Number the pictures 1 to 3, going from lightest to heaviest.

1.

______ ______ ______

2.

______ ______ ______

3.

______ ______ ______

4.

______ ______ ______

Which weighs more? Use the > (greater than) or < (less than) symbol.

5.

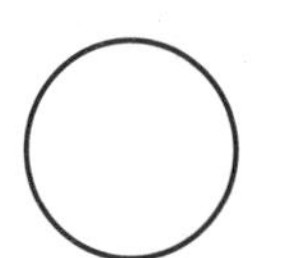

6.

7.

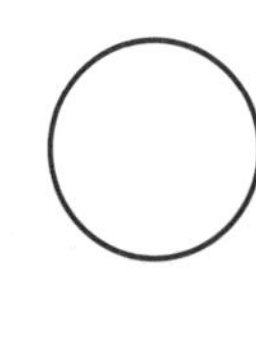

8.

Practice 2

Each section of the bar equals one inch. Measure each item to the nearest inch.

1.

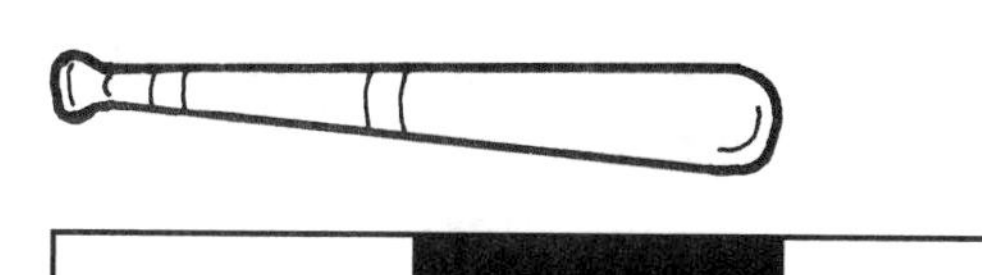

The bat is ________ inches long.

2.

The hat is ________ inch long.

3.

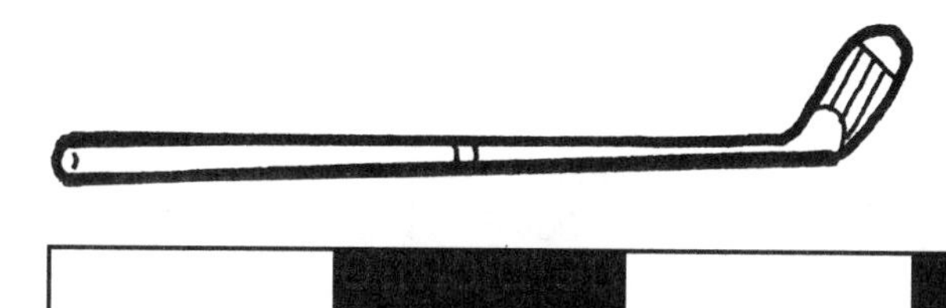

The golf club is ________ inches long.

4.

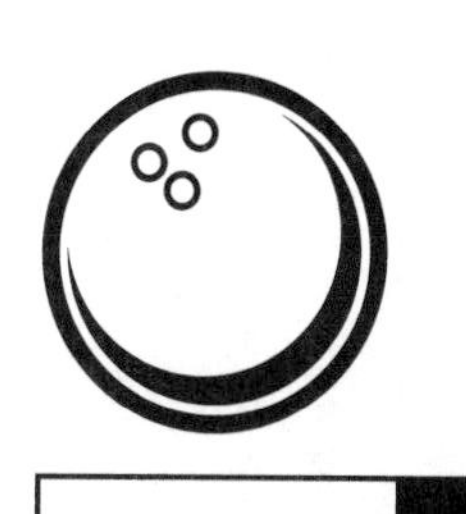

The bowling bowl is ________ inch long.

5.

The bowling pins are ________ inches long.

6.

The ski is ________ inches long.

Practice 3

Write the temperature on the line.

1.

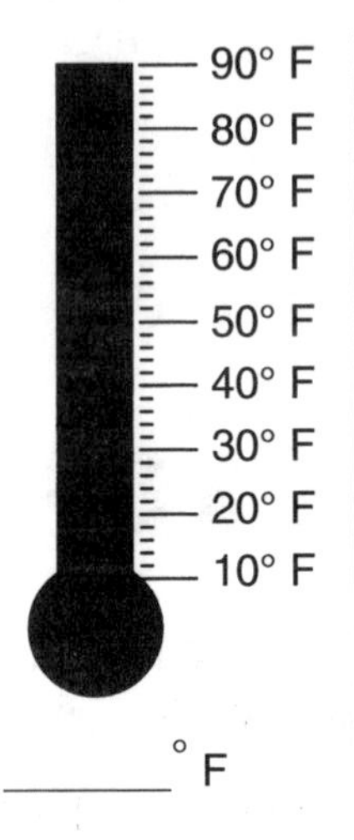

_______ ° F

2.

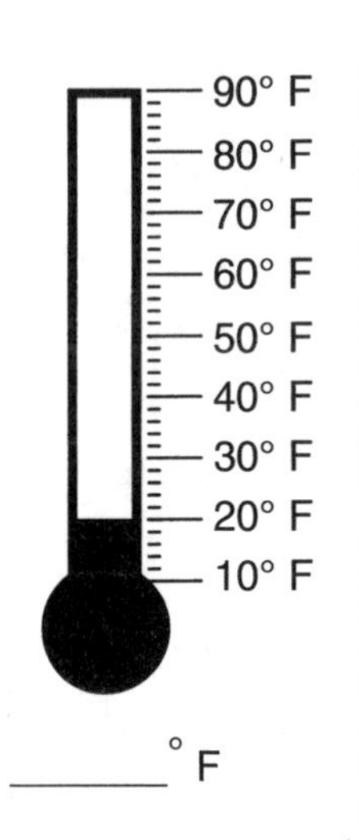

_______ ° F

3.

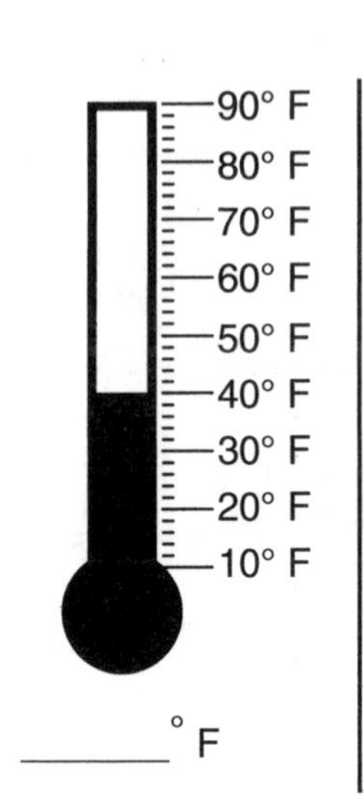

_______ ° F

4.

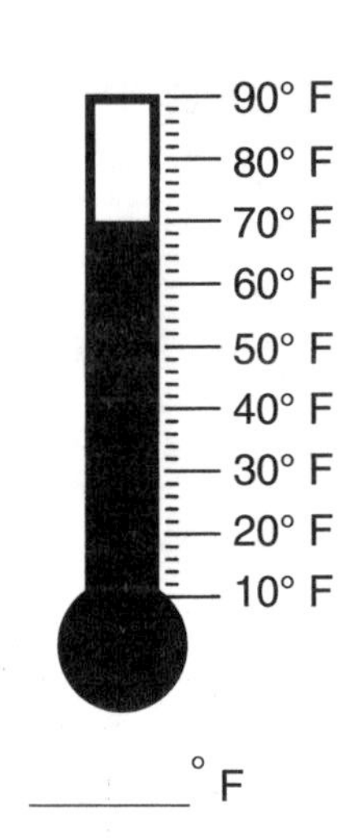

_______ ° F

Color the thermometer to show the temperature.

5.

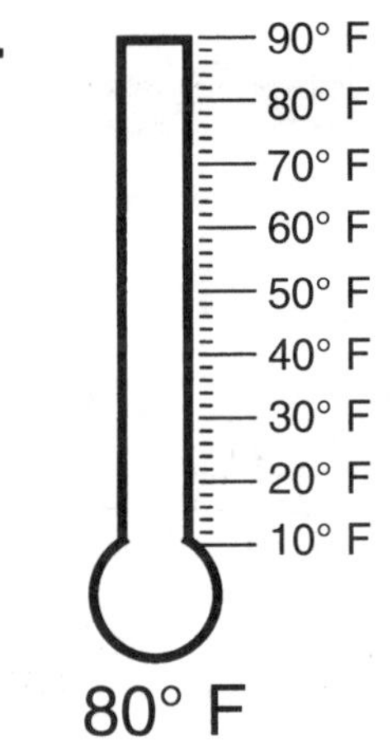

80° F

6.

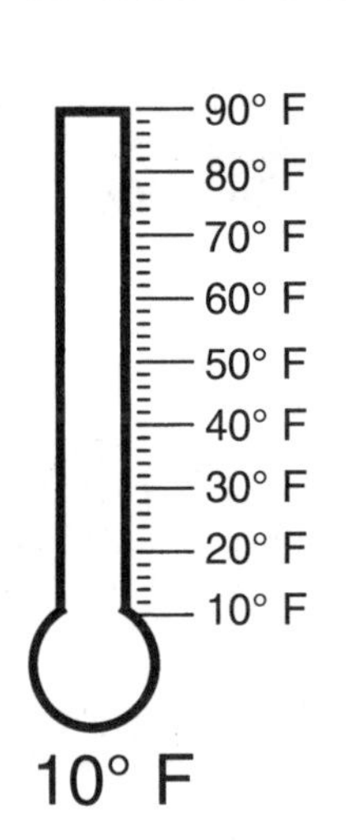

10° F

7.

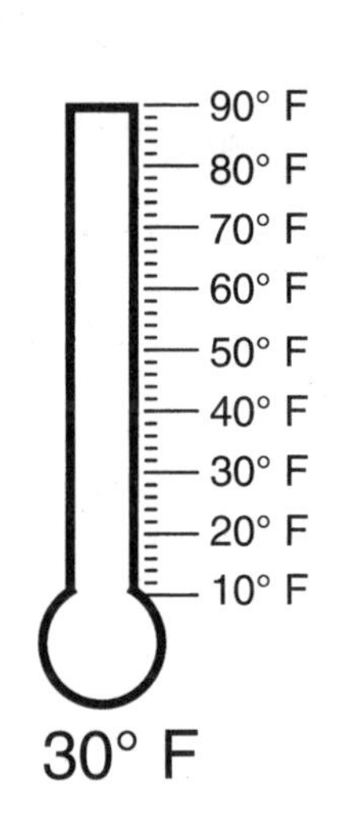

30° F

8.

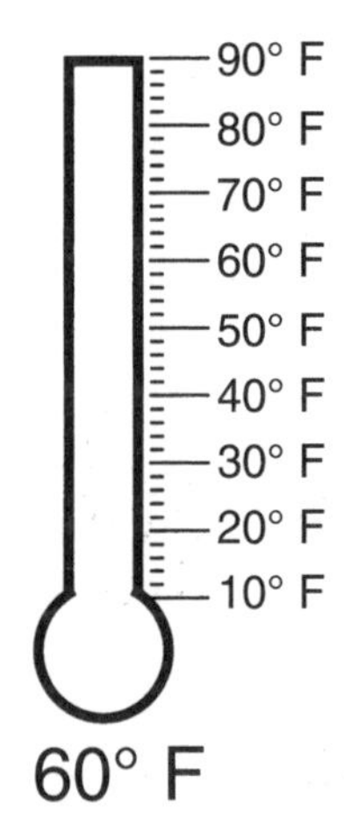

60° F

Match each picture to its temperature.

100° F

80° F

60° F

0° F

Practice 4

Solve the clue. Complete the crossword puzzle using the Answer Box.

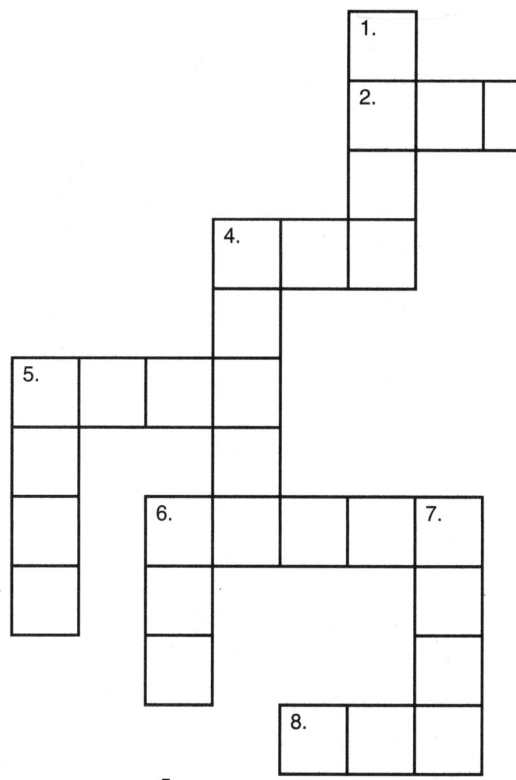

Answer Box

zero	one	two	three
four	five	six	seven
eight	nine	ten	

Across

2.  = __________ legs

4.   = __________ leaves

5. 5 – 1 = __________

6. 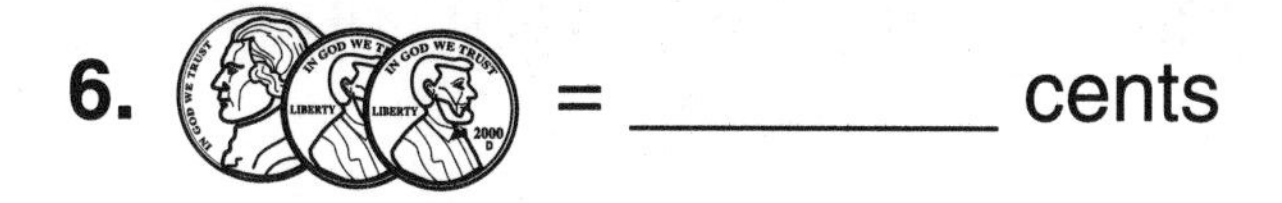 = __________ cents

8. 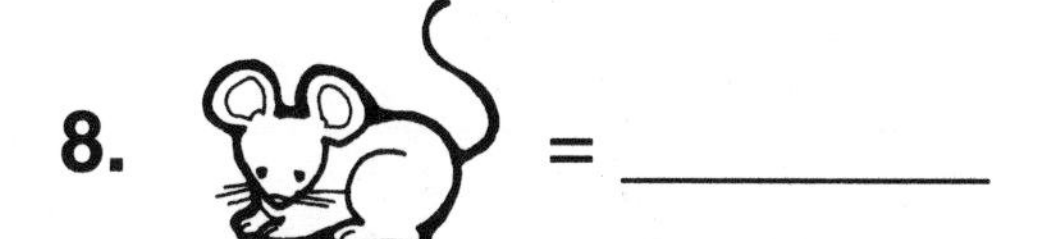 = __________

Down

1.  = __________

3. 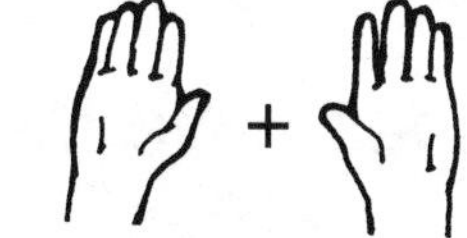 = ______ fingers

4. = ______ bears

5. **eight – three** = __________

6.   = __________ hats

7.  = __________ cents

Practice 5 

Count the number of square units to find the area.

A.

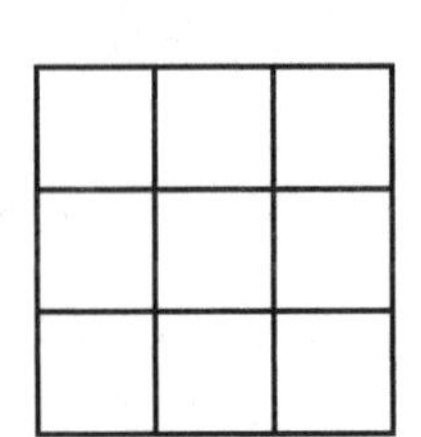

_____ square units

B.

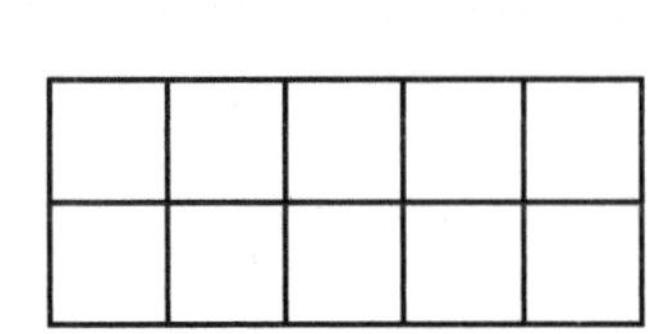

_____ square units

C.

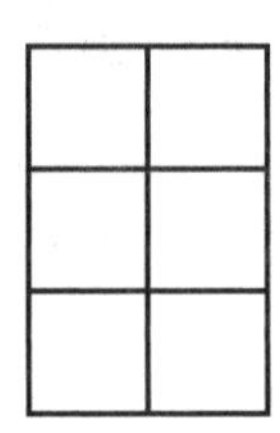

_____ square units

D.

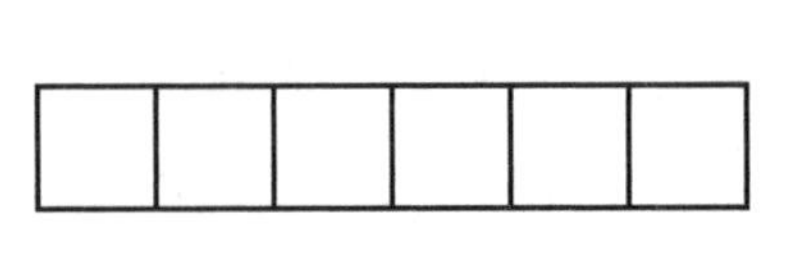

_____ square units

E.

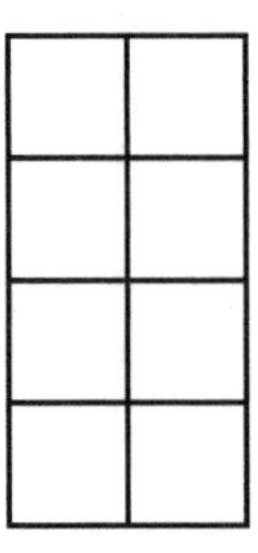

_____ square units

F.

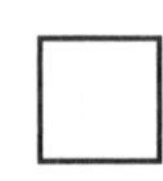

_____ square unit

Answer the questions.

1. Which shape has the most square units? ____________________

2. Which two shapes have the same number of square units?

____

3. Which shape has the fewest number of square units? __________

4. Draw a shape with 5 square units.	**5.** Draw a shape with 3 square units.	**6.** Draw a shape with 4 square units.

Practice 6

Color the items that belong in the family.

Shape Family	Choices
1.	
2.	
3.	
4.	
5.	

Color the plane (flat) and 3-dimensional shape that belong in the same family.

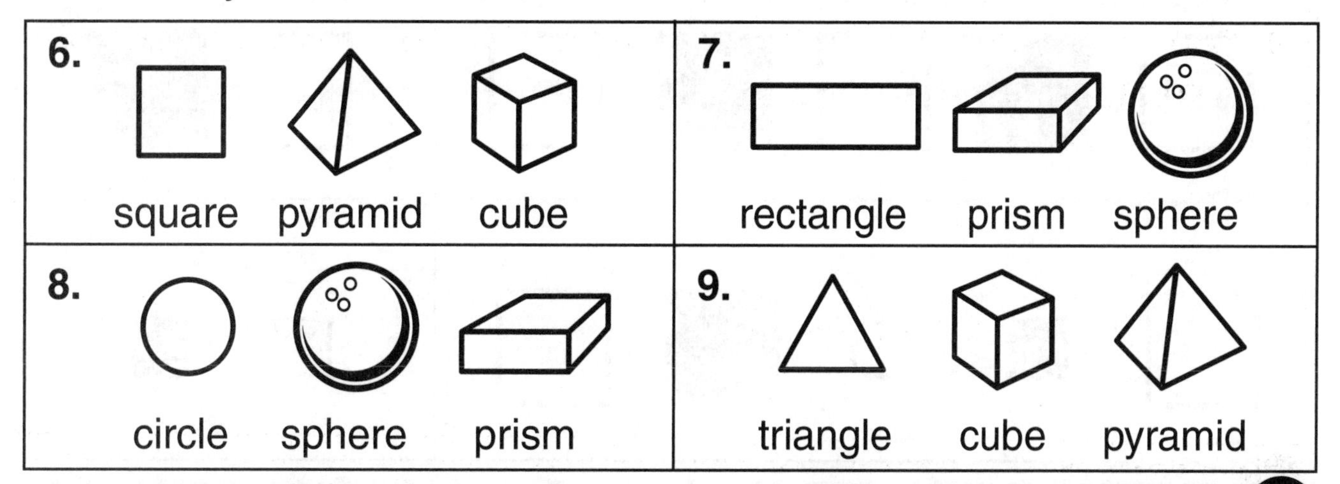

Practice 7

Color the item that should come next in the pattern.

Shape Patterns	Choices
1.	
2.	
3.	
4.	
5.	
6.	

Practice 8

Count the number of equal parts.

1.

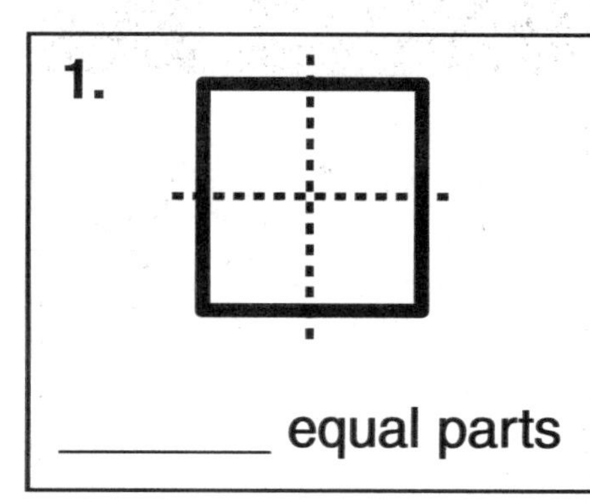

_______ equal parts

2.

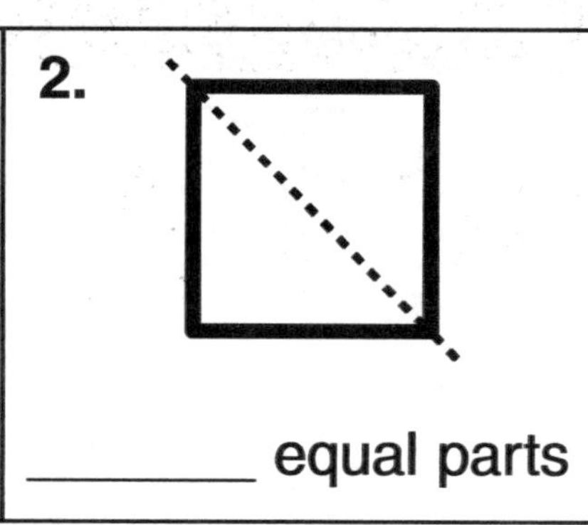

_______ equal parts

3.

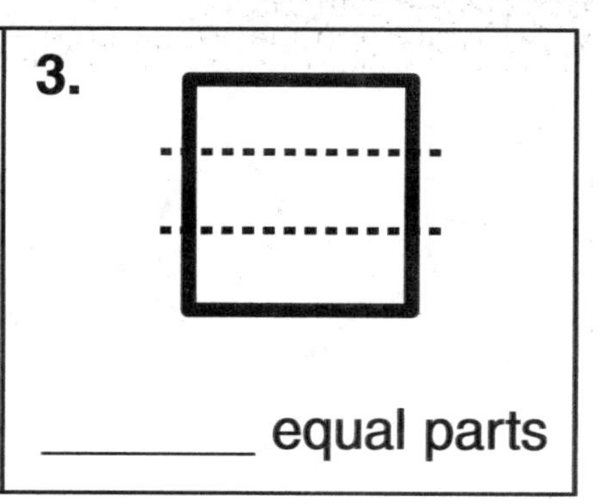

_______ equal parts

Is the shape symmetrical? Circle "yes" or "no."

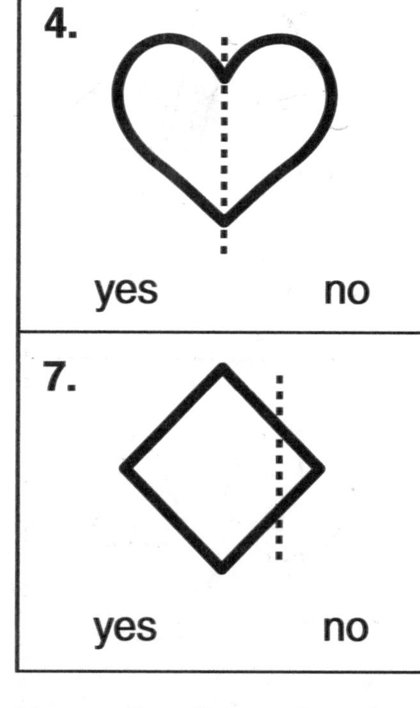

4.

yes no

5.

yes no

6.

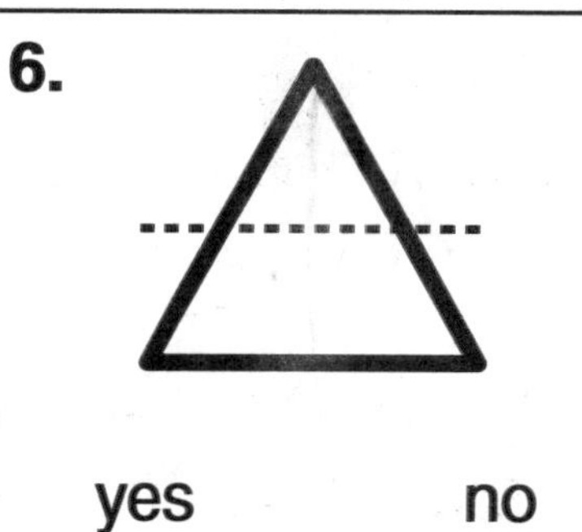

yes no

7.

yes no

8.

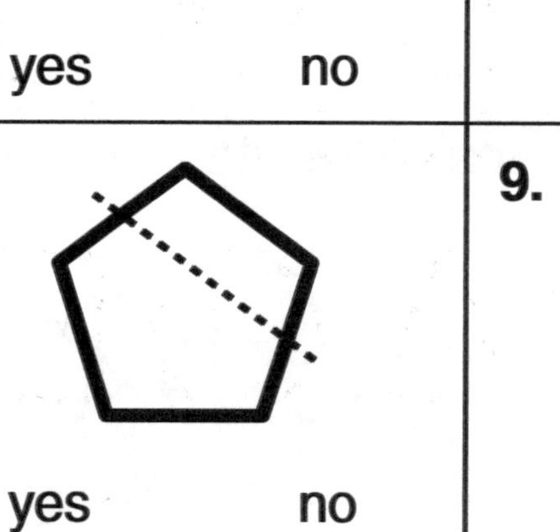

yes no

9.

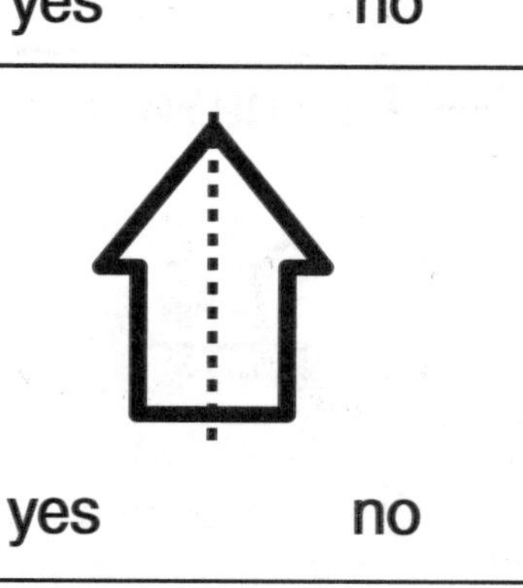

yes no

Draw the line of symmetry.

10. **11.** **12.**

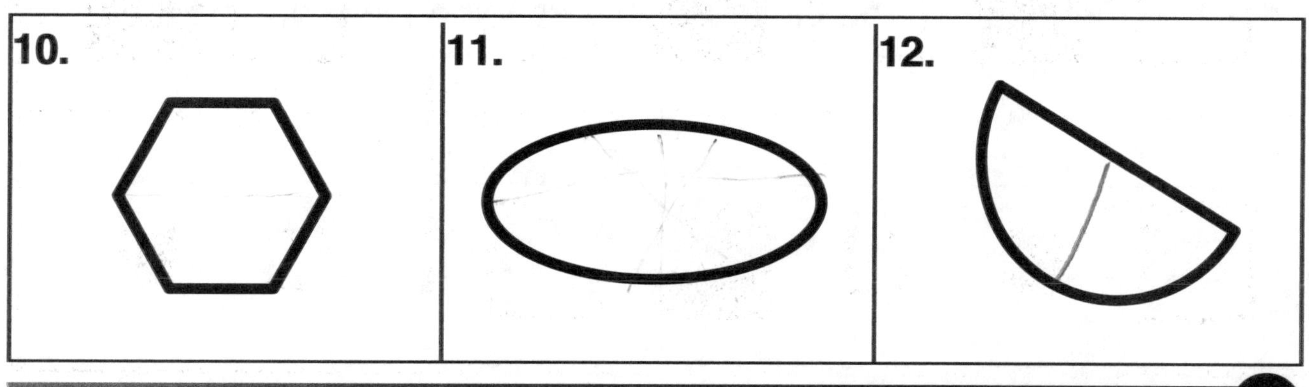

Practice 9

Fill in the answer circle under the correct picture for problems 1–3.

1. The circle is <u>in front of</u> the square

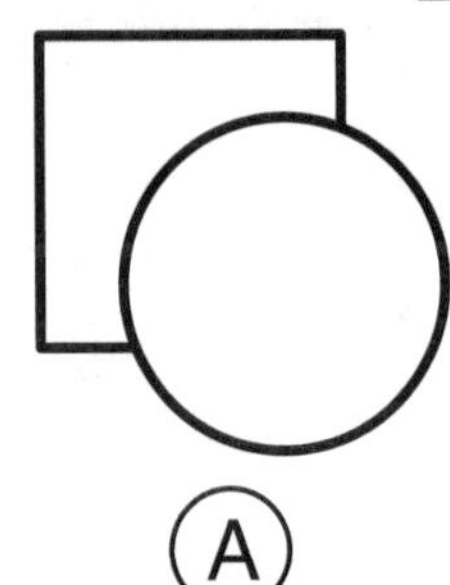

Ⓐ

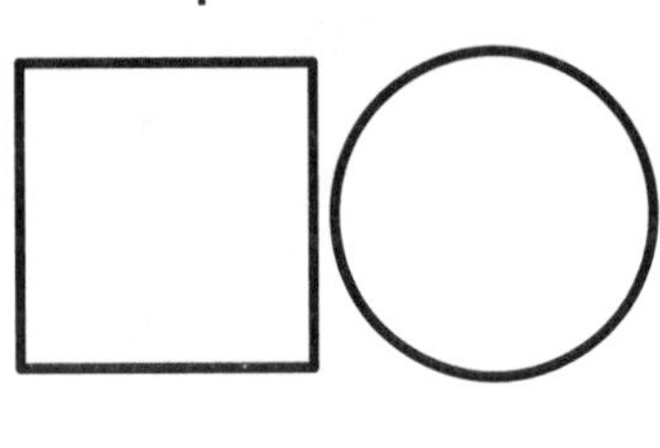

Ⓑ

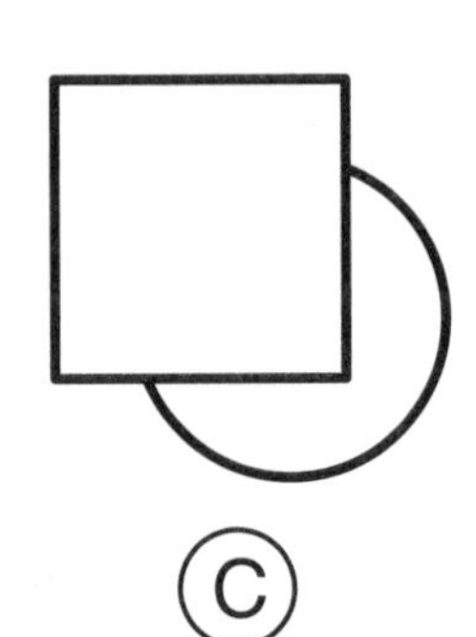

Ⓒ

2. The star is in <u>between</u> the circle and the square.

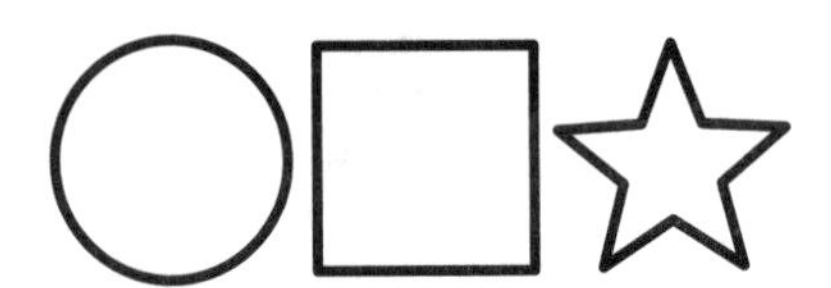

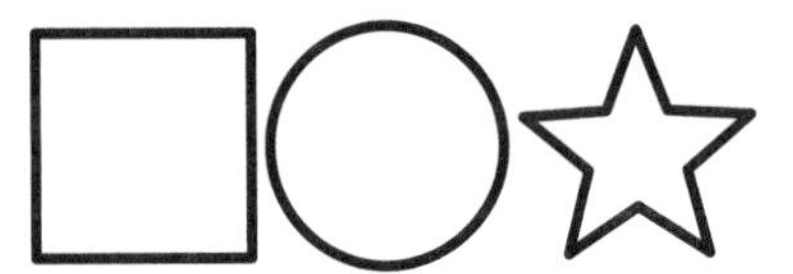

3. The square is <u>on top</u> of the star.

Ⓐ

Ⓑ

Ⓒ

4. Draw the picture. The circle is <u>on top</u> of the square.

5. Draw the picture. The star is <u>below</u> the circle.

Practice 10

Count the money and then write the price for each item on the line.

1.	2.	3.	4.
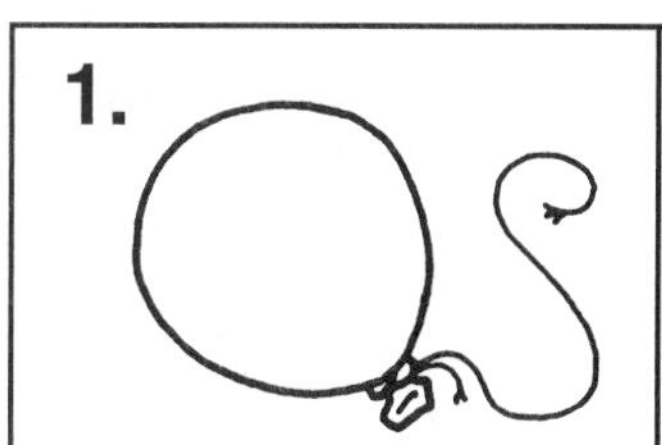			

______ ¢	______ ¢	______ ¢	______ ¢

Add.

5.	6.	7.
 2¢	 1¢	 2¢
+  + 3¢	+ 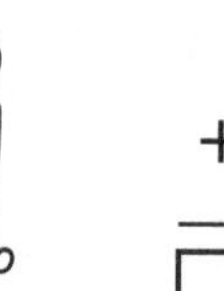 + 4¢	+ + 4¢
¢	¢	¢
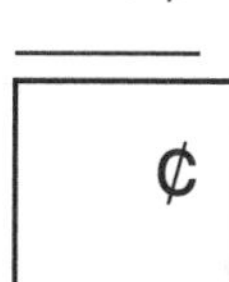		

8.	9.	10.
1¢	 4¢	2¢
 2¢	 3¢	 4¢
+ + 4¢	+ 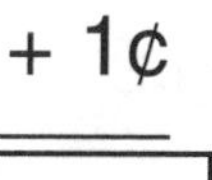 + 1¢	+ + 3¢
¢	¢	¢

Practice 11

Complete the table.

Number of Nickels	1	2	3	4	5	6	7	8	9	10
Amount of Money	5¢	____¢	____¢	20¢	____¢	____¢	____¢	40¢	____¢	50¢

Read and solve each problem. Circle the correct answer. Use the table above to help you.

1. Jerome has 2 pockets. In each pocket Jerome has 2 nickels. How much money does Jerome have?

10¢ 15¢ 20¢

2. C.J. has 1 pocket. In her pocket she has 5 nickels. How much money does C.J. have?

1¢ 5¢ 25¢

3. Mabel has 3 pockets. In each pocket Mabel has 1 nickel. How much money does Mabel have?

3¢ 5¢ 15¢

4. Neil has 35¢ in 1 pocket. How many nickels does Neil have?

3 5 7

5. Jane has 45¢ in 2 pockets. How many nickels does Jane have?

5 9 12

6. R.J. has 50¢ in 5 different pockets. How many nickels does R.J. have?

5 10 15

Practice 12          

Count the money.

1.  ______¢ = A	**2.**  ______¢ = I	**3.**  ______¢ = U
4.  ______¢ = N	**5.**  ______¢ = T	**6.**  ______¢ = !
7.  ______¢ = O	**8.**  ______¢ = S	**9.**  ______¢ = C

Write the letter on the line that matches each amount above.

___		___	___	___		___	___	___	___	___
1¢		7¢	6¢	2¢		7¢	5¢	8¢	2¢	9¢

___	___	___	___	___	___
7¢	5¢	1¢	2¢	3¢	4¢

Practice 13        

2¢	**3¢**	**4¢**	**5¢**	**6¢**

Write the subtraction sentence.

1. Dan has 8¢. He buys a (hat). How much does he have left?

____ ¢ – ____ ¢ = ____ ¢

2. Pam has 5¢. She buys a (pail). How much does she have left?

____ ¢ – ____ ¢ = ____ ¢

3. Jan has 6¢. She buys (sunglasses). How much does she have left?

____ ¢ – ____ ¢ = ____ ¢

4. Stella has 9¢. She buys a (pail). How much does she have left?

____ ¢ – ____ ¢ = ____ ¢

5. Sam has 4¢. He buys a (ball). How much does he have left?

____ ¢ – ____ ¢ = ____ ¢

6. Stan has 7¢. He buys a (shell). How much does he have left?

____ ¢ – ____ ¢ = ____ ¢

7. Jim has 7¢. He buys a (hat). How much does he have left?

____ ¢ – ____ ¢ = ____ ¢

8. Fran has 6¢. She buys a (shell). How much does she have left?

____ ¢ – ____ ¢ = ____ ¢

Practice 14          

Complete the table.

Number of Dimes	1	2	3	4	5	6	7	8	9
Amount of Money	10¢	____¢	____¢	40¢	____¢	____¢	70¢	____¢	90¢

Add or subtract to solve each word problem.

1. Herb has 20¢. He spent 10¢ buying a candy bar. How much money does Herb have left?

Herb has ______ ¢ left.

[ ] ¢
− [ ] ¢
[ ] ¢

2. Betty has 30¢ in one pocket and 40¢ in another pocket. How much money does Betty have?

Betty has ______ ¢.

[ ] ¢
+ [ ] ¢
[ ] ¢

3. Dale had 50¢. He found 10¢ in his pocket. How much money does Dale have?

Dale has ______ ¢.

[ ] ¢
+ [ ] ¢
[ ] ¢

4. Fay had 80¢. She spent 70¢ buying a CD. How much money does Fay have left?

Fay has ______ ¢ left.

[ ] ¢
− [ ] ¢
[ ] ¢

5. Sid had 90¢. He gave 60¢ to his brother. How much money does Sid have left?

Sid has ______ ¢ left.

[ ] ¢
− [ ] ¢
[ ] ¢

6. May had 0¢. She earned 50¢ selling lemonade. How much money does May now have?

May has ______ ¢.

[ ] ¢
+ [ ] ¢
[ ] ¢

Practice 15

Draw a line from each set of pennies to its equivalent amount in dimes.

1. 

2. 

3. 

4. 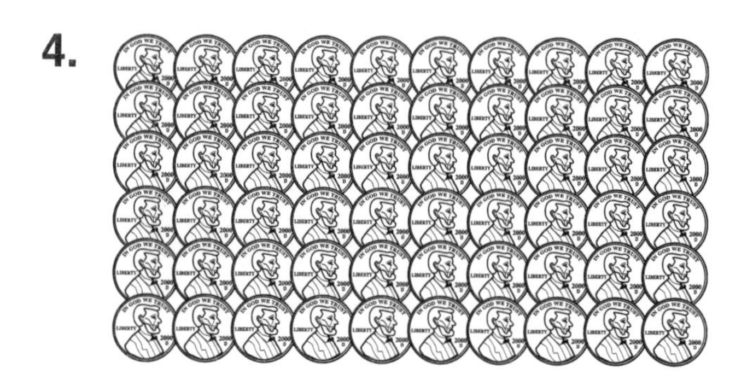

5. 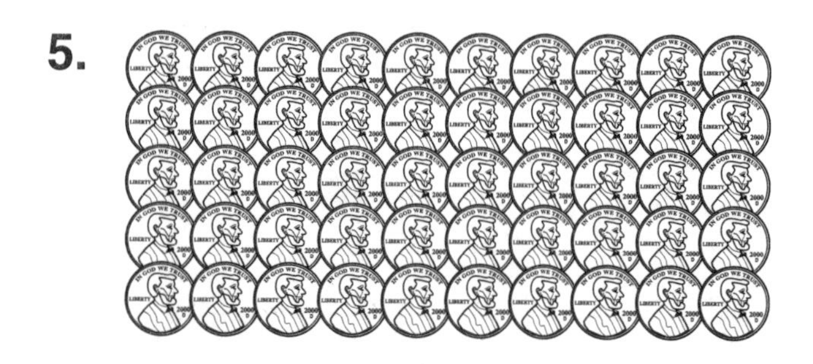

6. 

Practice 16            

Write the time on the lines.

1.

___ : ___ ___

2.

___ : ___ ___

3.

___ : ___ ___

4.

___ : ___ ___

5.

___ : ___ ___

6.

___ : ___ ___

Draw the hands to show the time.

7.

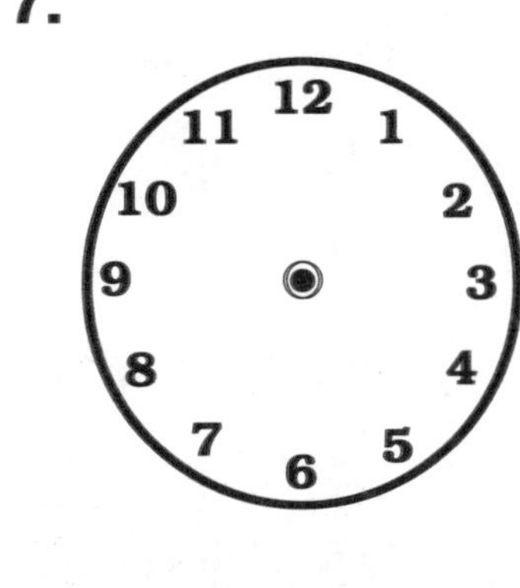

5:00

8.

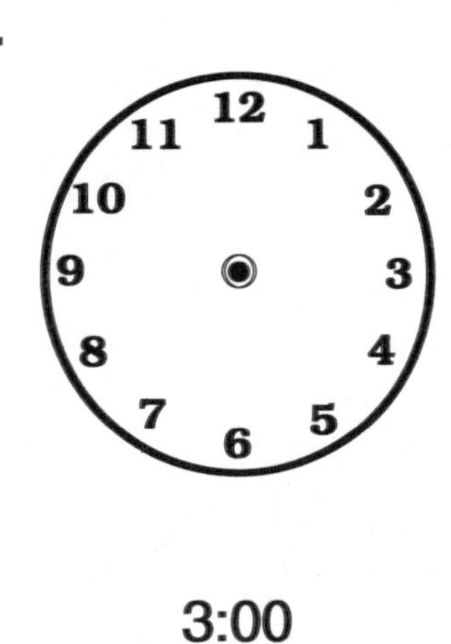

3:00

9.

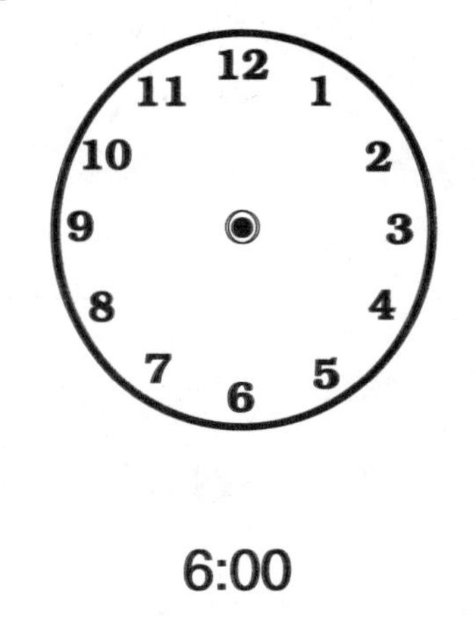

6:00

Practice 17 

Circle the answer.

1. Which would take more time? running a mile walking a mile	**2.** Which would take more time? walking to school driving to school	**3.** Which would take more time? washing windows washing a face
4. Which would take less time? counting $1 in dimes counting $1 in pennies	**5.** Which would take less time? writing a letter writing a postcard	**6.** Which would take less time? filling the sink filling the pool

Use numbers to rewrite the time.

7. one o'clock ___ : ___ ___	**8.** ten o'clock ___ : ___ ___	**9.** six o'clock ___ : ___ ___
10. four o'clock ___ : ___ ___	**11.** eight o'clock ___ : ___ ___	**12.** twelve o'clock ___ : ___ ___

Practice 18

Draw the hands to show half an hour later. Write the time on the line.

1.

3:00 ____ : ____ ____

2.

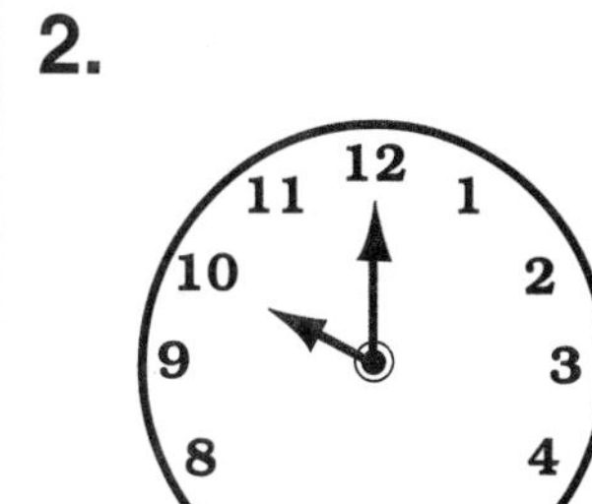

10:00 ____ : ____ ____

Read the problem. Draw the hands on the clock to show the time. Write the time on the line.

3. The piano lesson began at 2:00 and lasted half an hour. What time did the lesson end?

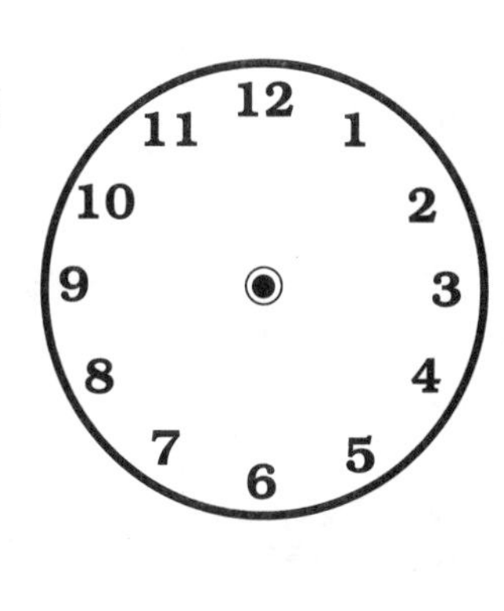

____ : ____ ____

4. Band practice began at 4:00 and lasted half an hour. What time did band practice end?

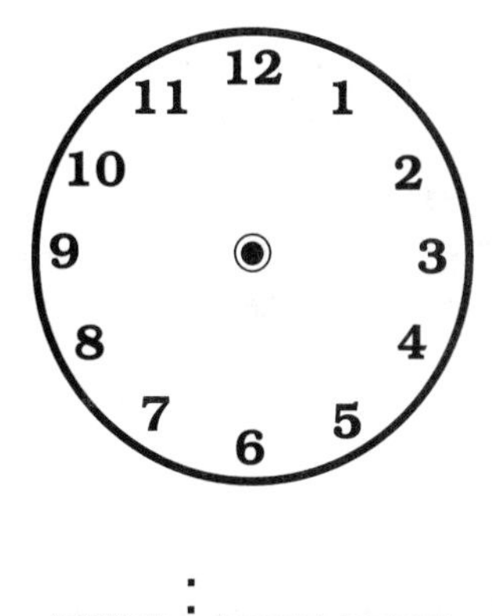

____ : ____ ____

5. The violin lesson began at 1:00 and lasted half an hour. What time did the lesson end?

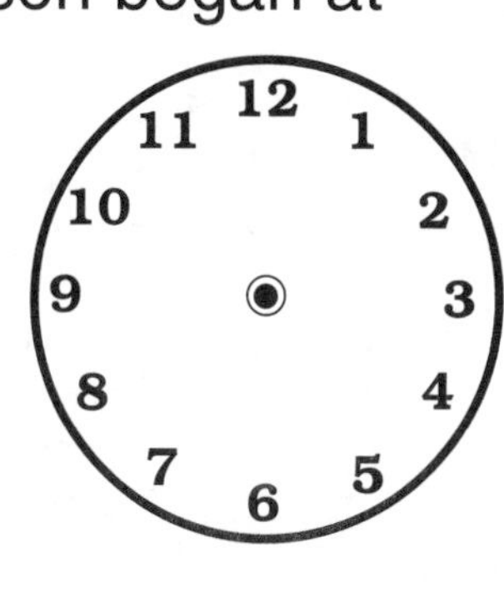

____ : ____ ____

6. The singing lesson began at 12:00 and lasted half an hour. What time did the lesson end?

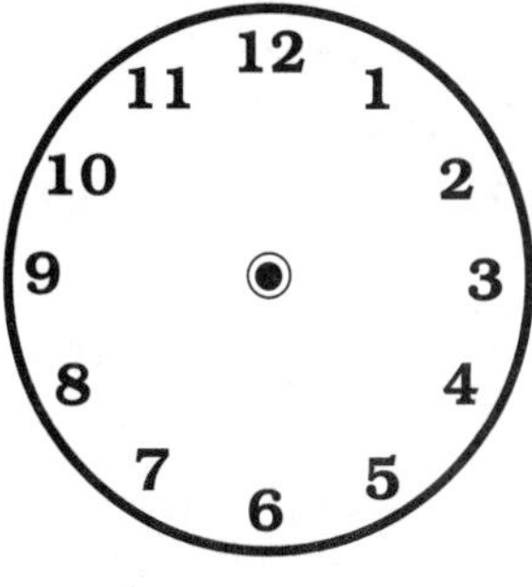

____ : ____ ____

Practice 19

Count the number of animals. Circle the larger number.

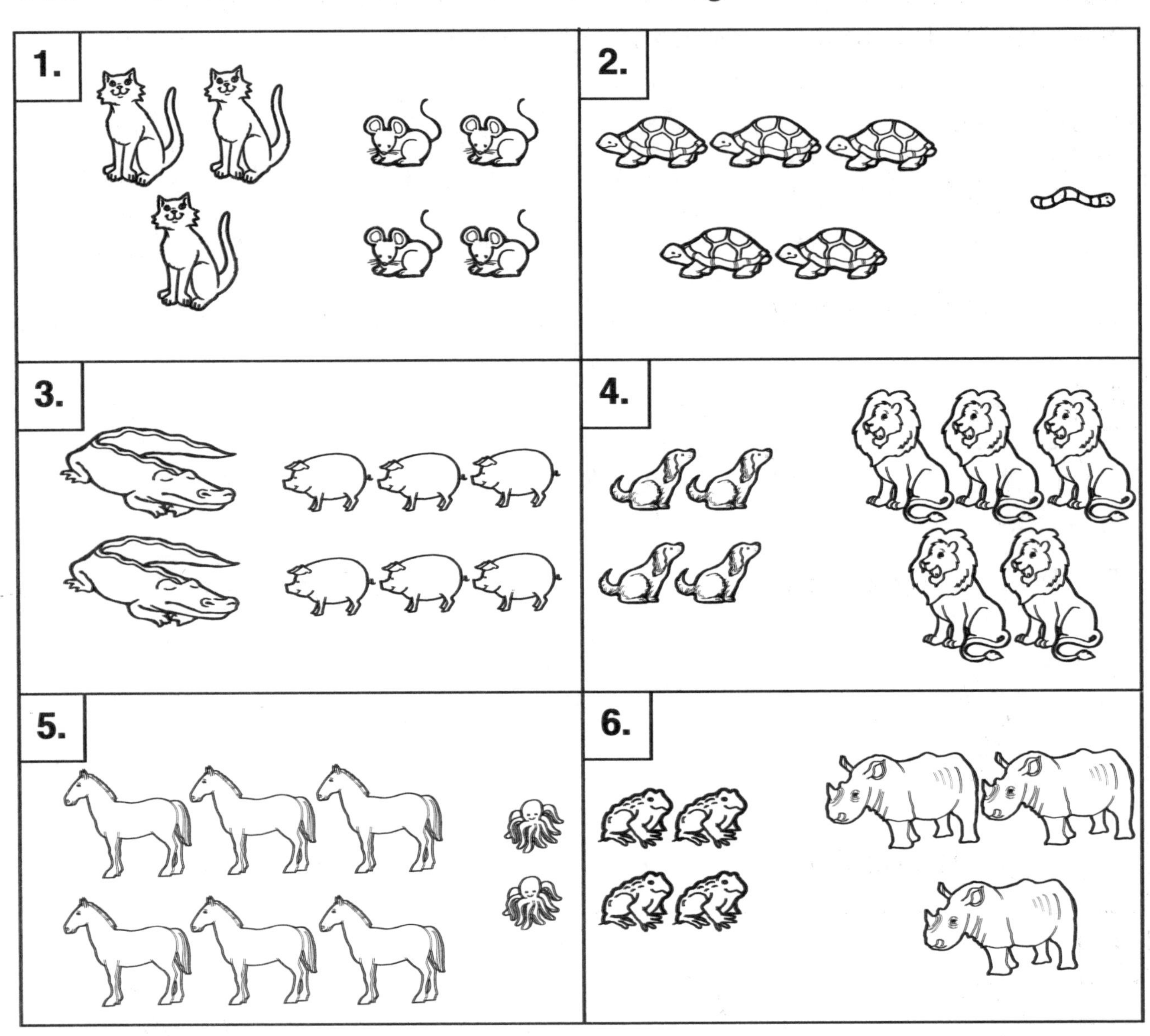

Who has more pets? Circle the name.

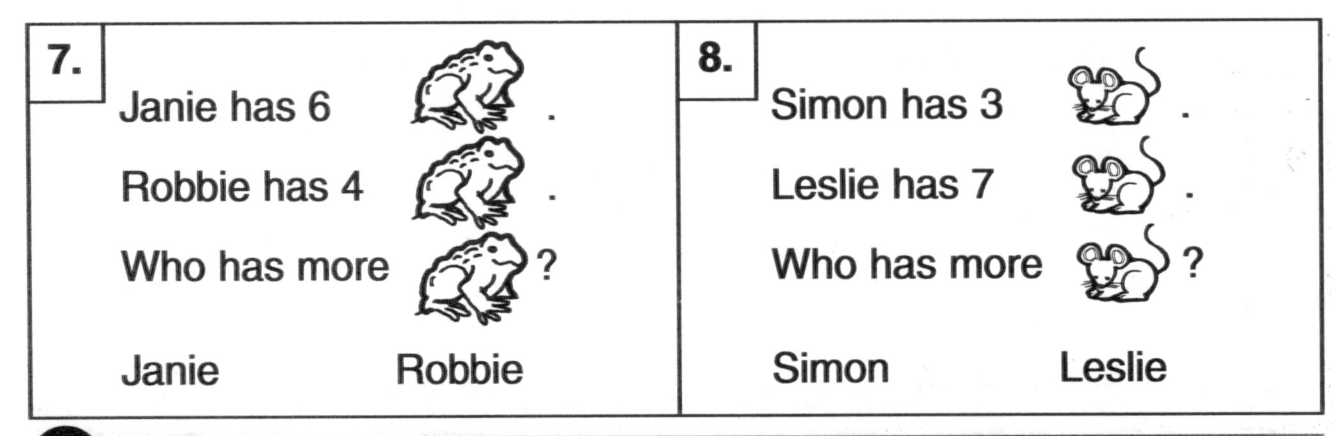

Practice 20

Use the < (less than), > (greater than), or = (equal to) symbol to compare the numbers. Complete each sentence.

1.

10   ◯ 5 

_______ is greater than _______ .

2.

9  ◯ 7 

_______ is greater than _______ .

3.

7 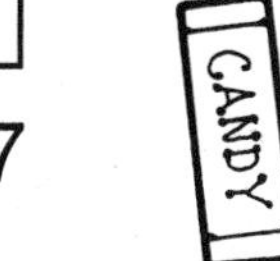

◯ 10 

_______ is less than _______ .

4.

3 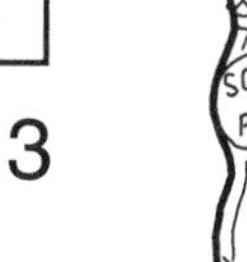

◯ 2 

_______ is greater than _______ .

5.

8  ◯ 8 

_______ is equal to _______ .

6.

6 

◯ 1 

_______ is greater than _______ .

7.

4  ◯ 3 

_______ is greater than _______ .

8.

2  ◯ 9 

_______ is less than _______ .

9.

4   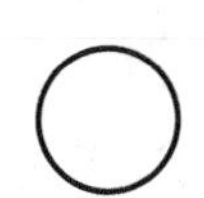 ◯ 7 

_______ is less than _______ .

10.

7   ◯ 7 

_______ is equal to _______ .

Practice 21

Draw a line under the smallest number, and circle the largest number. Then write the numbers in order, smallest to largest.

1.

Dave has 7 action figures.

Steve has 2 action figures.

Brett has 8 action figures.

_____, _____, _____

2.

Pachai has 1 bracelet.

Chelsea has 7 bracelets.

Briana has 3 bracelets.

_____, _____, _____

3.

Alex has 10 pencils.

Luis has 5 pencils.

Frankie has 1 pencil.

_____, _____, _____

4.

Tamra has 3 pails.

Karen has 6 pails.

Maria has 9 pails.

_____, _____, _____

5.

Michael has 3 books.

Bryan has 4 books.

Eric has 0 books.

_____, _____, _____

6.

Gayle has 6 hats.

Jennifer has 8 hats.

Mavis has 10 hats.

_____, _____, _____

Practice 22             

Complete each table by adding the top number to each number below.

Example

+ 2	
5	7
1	3
10	12
9	11

1.

+ 5	
0	
2	
4	
1	

2.

+ 3	
6	
1	
2	
4	

3.

+ 6	
4	
3	
1	
2	

4.

+ 4	
5	
6	
4	
3	

5.

+ 2	
7	
8	
6	
3	

6.

+ 0	
7	
9	
10	
8	

Write the rule.

7.

+ _____	
4	7
5	8
6	9

8.

+ _____	
1	2
3	4
8	9

9.

+ _____	
9	9
8	8
7	7

10.

+ _____	
6	10
1	5
3	7

Practice 23

Use the number line to practice counting on from a given number. Write the answer on the line.

1.

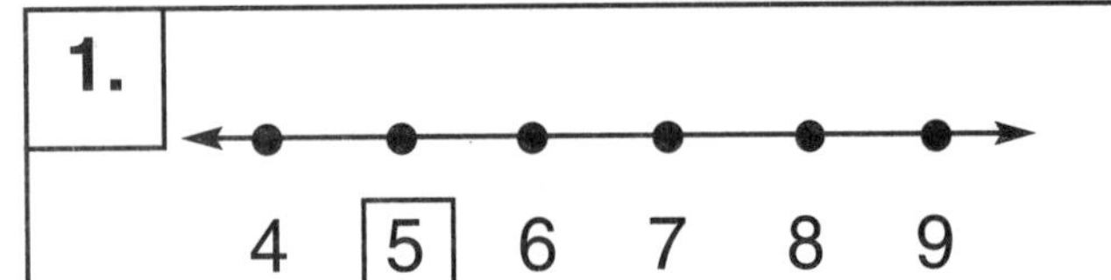

Jane has 5 flowers. Paula has 3 more flowers than Jane. How many flowers does Paula have?

Paula has _______ flowers.

2.

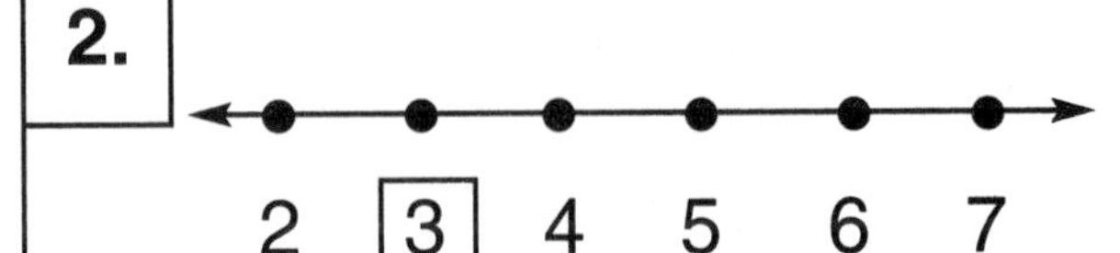

Adam has 3 toy cars. Jack has 4 more toy cars than Adam. How many toy cars does Adam have?

Adam has _______ toy cars.

3.

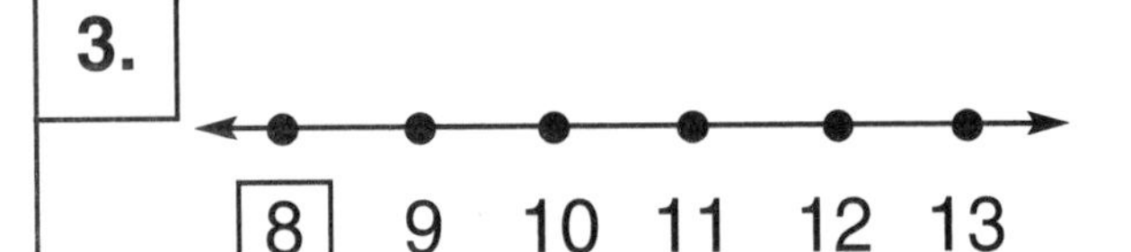

Beth caught 8 butterflies. Mason caught 5 more butterflies than Beth. How many butterflies did Mason catch?

Mason caught _______ butterflies.

4.

11 [12] 13 14 15 16

Patrick found 12 dinosaur bones. Melanie found 4 more bones than Patrick. How many bones did Melanie find?

Melanie found _________ bones.

Count by 2s. Color those numbers red.

1	2	3	4	5	6	7	8	9	10
11	12	13	14	15	16	17	18	19	20
21	22	23	24	25	26	27	28	29	30
31	32	33	34	35	36	37	38	39	40
41	42	43	44	45	46	47	48	49	50

Practice 24

Use the target board below to solve the problems.

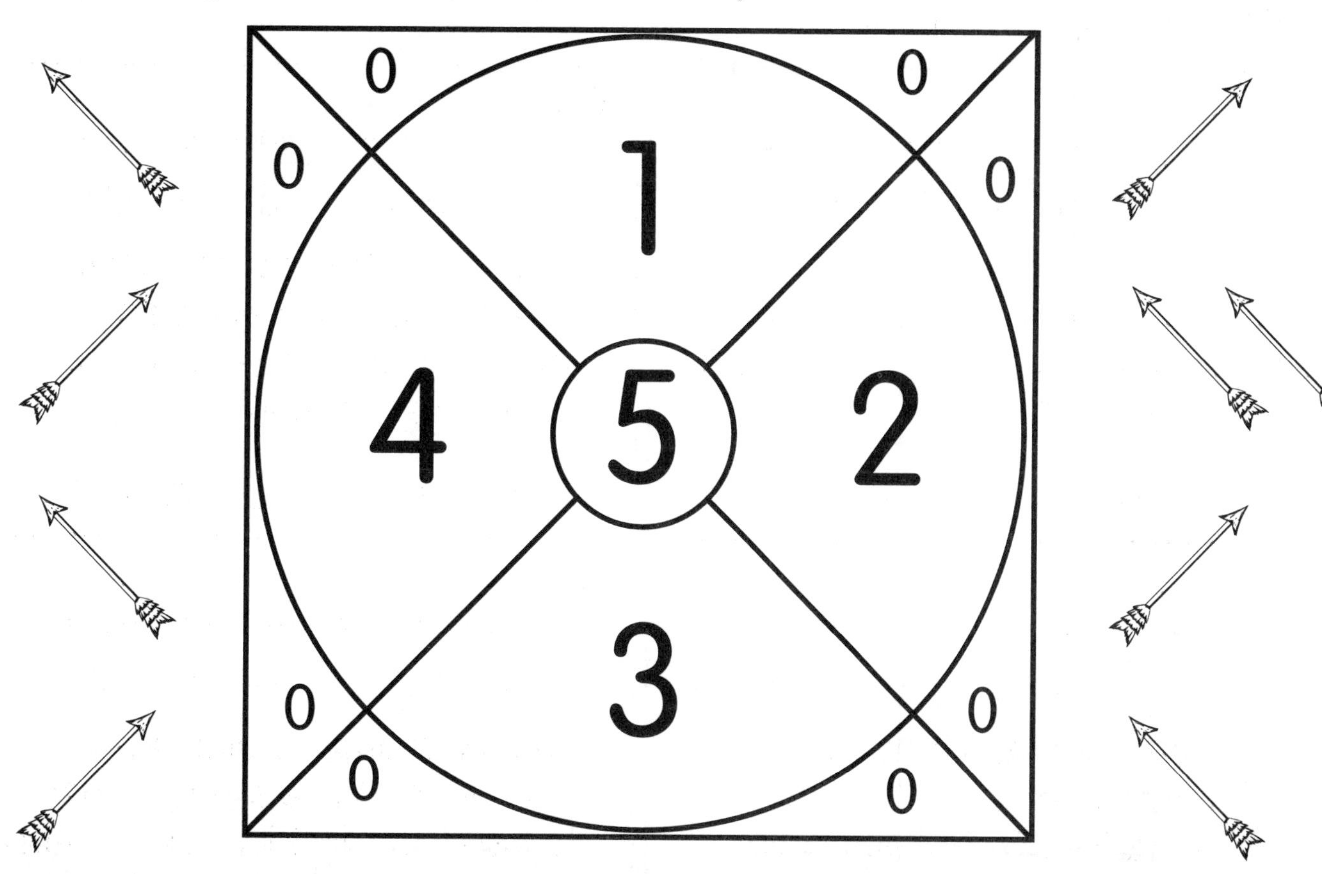

1. Julie shot two arrows for a total of 8 points. The first arrow hit a 5. What was the other number that Julie hit?

Julie hit ________________.

2. Keith shot two arrows for a total of 4 points. Both arrows hit the same number. What were the numbers Keith hit?

Keith hit a _____ and a _____.

3. Pete shot two arrows for a total of 5 points. The first arrow hit a 4. What was the other number that Pete hit?

Pete hit ________________.

4. Sally shot two arrows for a total of 2 points. What numbers did Sally hit?

Sally hit a _____ and a _____.

Practice 25            

Complete each table by subtracting the top number from each number below it.

Example

– 2	
2	0
10	8
7	5
5	3

1.

– 6	
10	
8	
7	
9	

2.

– 4	
6	
7	
9	
4	

3.

– 7	
9	
7	
10	
8	

4.

– 2	
10	
5	
6	
9	

5.

– 3	
4	
10	
5	
8	

6.

– 0	
8	
6	
10	
3	

Write the rule.

7.

– ____	
7	2
6	1
8	3

8.

– ____	
5	4
2	1
4	3

9.

– ____	
10	8
6	4
3	1

10.

– ____	
6	3
4	1
10	7

Practice 26

Circle the math sentences that equal the number at the top.

5
5 + 0
4 + 3
1 + 4
0 + 5

7
3 + 4
7 + 0
6 + 1
3 + 3

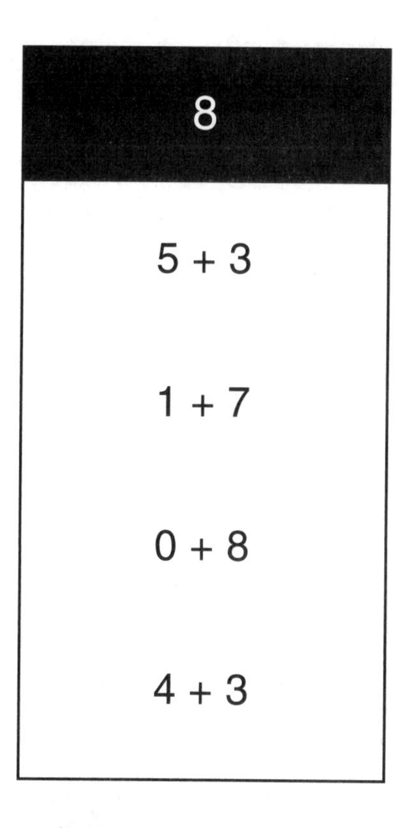

8
5 + 3
1 + 7
0 + 8
4 + 3

Circle the math sentences that equal the number at the top.

4
10 – 6
9 – 2
7 – 3
4 – 0

6
6 – 0
5 – 3
10 – 4
3 – 3

2
5 – 1
2 – 0
4 – 2
9 – 7

Practice 27          

Read each word problem. Circle the better estimate.

1. Tina has a handful of pennies. About how many pennies does Tina have?

7 pennies 70 pennies

2. Steve has a bowl with a goldfish. About how many goldfish does Steve have?

1 goldfish 0 goldfish

3. Bill has a bag full of many marbles. About how many marbles does Bill have?

2 marbles 20 marbles

4. Tilly has lost some teeth. About how many teeth has Tilly lost?

4 teeth 40 teeth

5. About how long should it take to brush your teeth?

3 seconds 3 minutes 3 hours

6. About how long does it take a cake to bake?

1 second 1 minute 1 hour

Practice 28

Fill in the circle under the number that completes each math sentence.

1. 6 + _______ = 10

3 ○ 4 ○ 5 ○ 6 ○

2. 2 + _______ = 9

6 ○ 7 ○ 8 ○ 9 ○

3. 10 + _______ = 10

0 ○ 1 ○ 2 ○ 3 ○

4. 1 + _______ = 5

1 ○ 2 ○ 3 ○ 4 ○

5. 8 + _______ = 10

0 ○ 1 ○ 2 ○ 3 ○

6. 2 + _______ = 8

3 ○ 4 ○ 5 ○ 6 ○

7. 10 – _______ = 7

0 ○ 1 ○ 2 ○ 3 ○

8. 10 – _______ = 1

6 ○ 7 ○ 8 ○ 9 ○

9. 7 – _______ = 2

5 ○ 6 ○ 7 ○ 8 ○

10. 6 – _______ = 1

3 ○ 4 ○ 5 ○ 6 ○

11. 7 – _______ = 4

3 ○ 4 ○ 5 ○ 6 ○

12. 9 – _______ = 4

4 ○ 5 ○ 6 ○ 7 ○

Practice 29

Finish numbering the bears in order. Then follow the directions in the box below.

A.  **B.**  **C.**  **D.** **E.** 

___1st___ _____nd _____rd _____th _____th

- Draw a  on the 3rd bear.
- Draw a  above the 5th bear's head.
- Draw a ○ around the 1st bear.
- Draw  on the 4th bear's face.
- Draw a  around the 2nd bear.

Follow the directions.

1. Circle the 3rd animal.

2. Circle the last animal.

3. Circle the 6th animal.

Practice 30

Answer the questions.

1. Who lives in the 5th apartment? ____________________
2. Which apartment is for rent? ____________________
3. How many children live in the 2nd apartment? ____________________
4. In which apartment does a baby live? ____________________
5. Which apartment is in between the 1st and the 3rd apartments?

   ____

6. How many children live in the apartments? ____________________

Follow the directions.

7. Circle the 7th heart.

8. Circle the 2nd star.

9. Circle the 10th arrow.

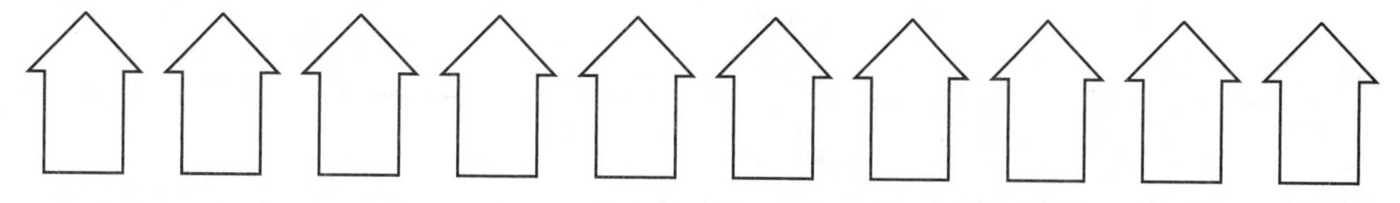

Practice 31

Complete the table by adding the missing numbers.

1		3	4		6		8	9	10
	12	13		15		17	18	19	
21	22		24	25	26				30
	32		34		36	37			40
41			44		46		48	49	

Write each number's "neighbors."

Example	to the left ←	to the right →	above ↑	below ↓
24	23	25	14	34
1. 33	______	______	______	______
2. 16	______	______	______	______
3. 37	______	______	______	______
4. 19	______	______	______	______

Practice 32

Complete the chart by adding the missing numbers.

51		53		55		57		59	60
61	62			65	66		68	69	
71			74		76			79	
	82	83			86	87		89	90
	92		94	95		97		99	

1. Color the numbers with a 0 in the ones place red.
2. Color the numbers with a 5 in the tens place orange.
3. Draw a yellow circle around the numbers with a 6 in the tens place.
4. Draw a green square around the numbers with the same digit in the tens and the ones places.
5. Draw a blue star on the numbers with a 3 in the ones place.
6. Draw a purple line under the numbers that have a 7 in the tens or ones places.

Complete each math sentence.

7. ______ > 86	**8.** ______ < 74	**9.** ______ > 51
10. ______ = 91	**11.** ______ < 60	**12.** ______ < 65

Practice 33

Use the hundreds chart to solve each riddle.

1	2	3	4	5	6	7	8	9	10
11	12	13	14	15	16	17	18	19	20
21	22	23	24	25	26	27	28	29	30
31	32	33	34	35	36	37	38	39	40
41	42	43	44	45	46	47	48	49	50
51	52	53	54	55	56	57	58	59	60
61	62	63	64	65	66	67	68	69	70
71	72	73	74	75	76	77	78	79	80
81	82	83	84	85	86	87	88	89	90
91	92	93	94	95	96	97	98	99	100

Riddle 1

1. I am larger than 30 and less than 50.
2. I am an even number.
3. When you count by 10s you say my name.

What number am I? ________

Riddle 2

1. I am less than 40 but larger than 10.
2. I have two numbers that are the same. My two numbers added together equal 4.

What number am I? ________

Riddle 3

1. I am larger than 50 and less than 100.
2. I have a 5 in the ones place.
3. I have a number smaller than 6 in the tens place.

What number am I? ________

Riddle 4

1. I have a 2 as one of my numbers.
2. Counting by 10s you say my name.

What number am I? ________

Practice 34

Circle sets of ten. Write the number of tens and ones.

1. 

_____ tens _____ ones

2. 

_____ tens _____ ones

3. 

_____ tens _____ ones

4. 

_____ tens _____ ones

5. 

_____ tens _____ ones

6. 

_____ tens _____ ones

Practice 35

Write the number of tens and ones, and then write the number.

1.

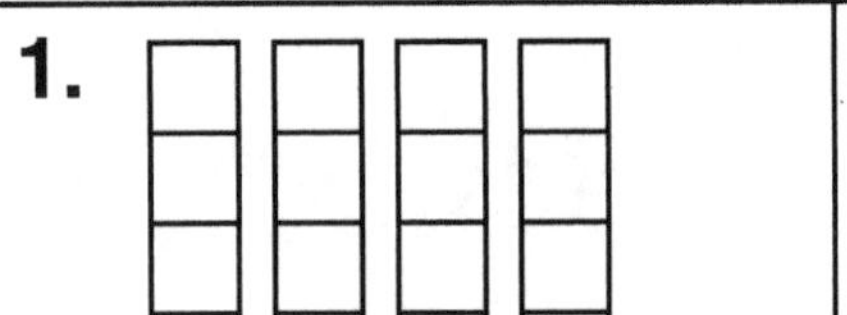

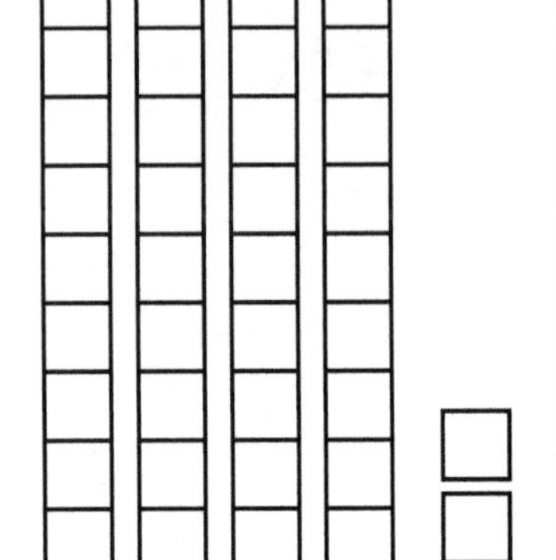

_____ tens _____ ones

Number = _____

2.

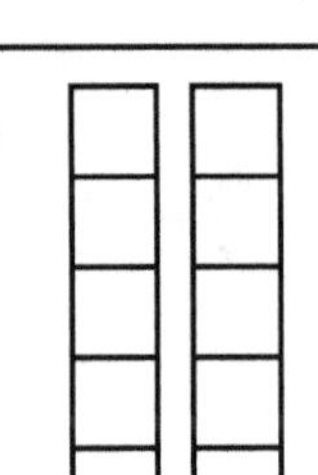

_____ tens _____ ones

Number = _____

3.

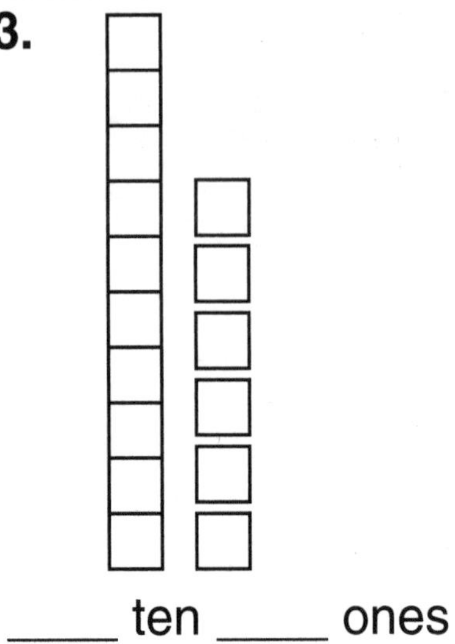

_____ ten _____ ones

Number = _____

4.

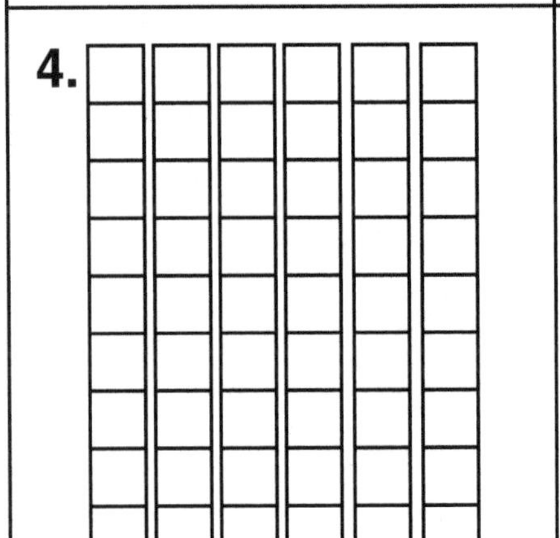

_____ tens _____ ones

Number = _____

5.

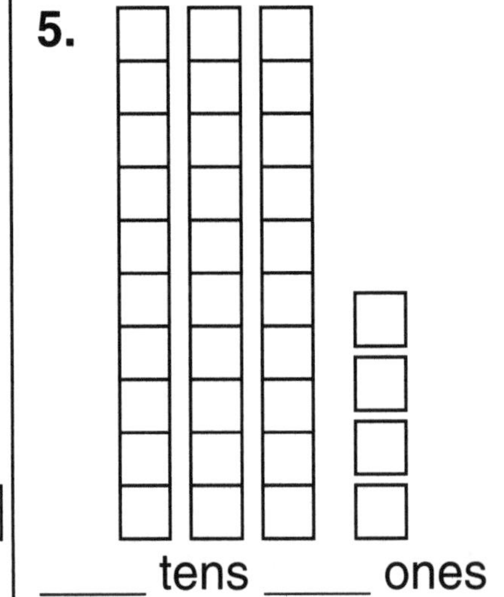

_____ tens _____ ones

Number = _____

6.

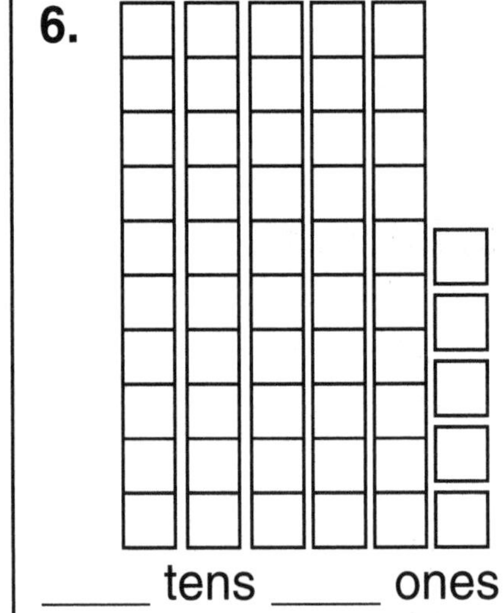

_____ tens _____ ones

Number = _____

7. Draw the tens and ones for the number 23.

8. Draw the tens and ones for the number 38.

Practice 36

Read and solve each word problem.

1. Lucas planted a seed. It grew 1 inch every day for a week. How tall is the plant? The plant is ______ inches tall.	**2.** Betsy is 38 inches tall. Her brother is 6 inches shorter. How tall is her brother? Betsy's brother is ____ inches tall.
3. Theodore's snake was 15 inches long. He grew 1 inch a month for the last 3 months. How long is Theodore's snake? Theodore's snake is ______ inches long.	**4.** Jason had a piece of yarn. He cut it into 2 pieces. Each piece was 6 inches long. How long was the yarn before Jason cut it? The yarn was ______ inches long.
5. Penny's zinnias are 5 inches tall. The tulips are twice as tall as the zinnias. How tall are the tulips? The tulips are ______ inches long.	**6.** Last year Andrew grew 4 inches. This year he is 35 inches tall. How tall was Andrew last year? Andrew was ______ inches tall.
7. Lacy is 25 inches tall. Barry is 3 inches taller than Lacy. Donna is 2 inches taller than Barry. How tall is Donna? Donna is ______ inches tall.	**8.** The spaghetti was 9 inches long. Rosa cut the spaghetti into 3 pieces of equal size. How long was each piece? Each piece was ______ inches long.

Test Practice 1        

Fill in the correct answer bubble.

1. Fill in the circle under the heaviest animal.

Ⓐ Ⓑ Ⓒ

2. Fill in the circle under the lightest animal.

Ⓐ Ⓑ Ⓒ

3. Measure the arrow. Fill in the circle under the correct answer.

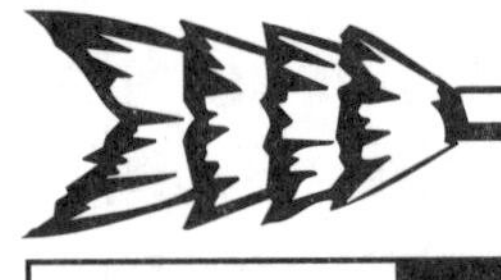 

4 inches	5 inches	6 inches
Ⓐ	Ⓑ	Ⓒ

4. Fill in the circle under the temperature.

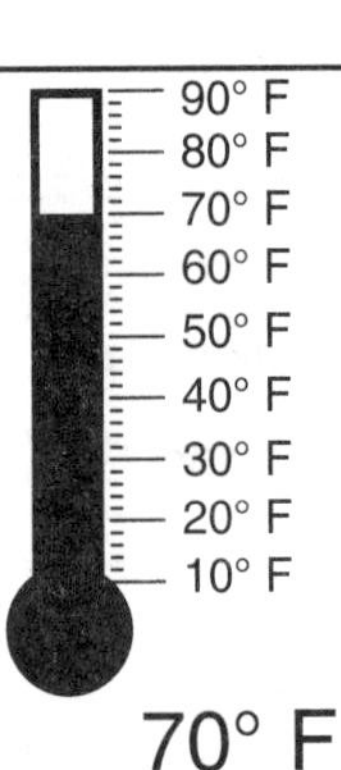

40° F	80° F	70° F
Ⓐ	Ⓑ	Ⓒ

5. Fill in the circle under the correct symbol.

5 ◯ 2

>	<	=
Ⓐ	Ⓑ	Ⓒ

6. Fill in the circle under the area.

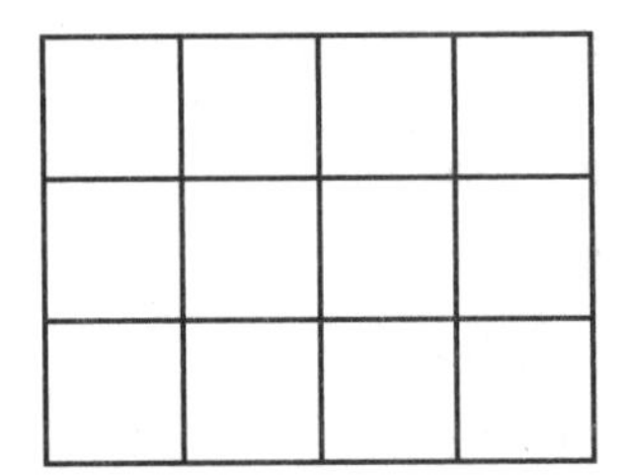

4 square units	8 square units	12 square units
Ⓐ	Ⓑ	Ⓒ

Test Practice 2

Fill in the correct answer bubble.

1. What shapes complete the pattern?

 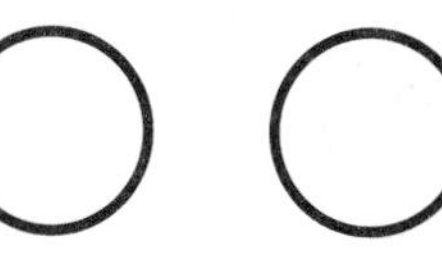  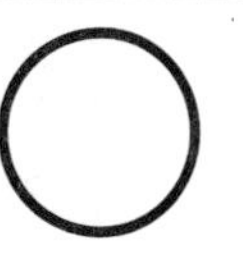

? ?

   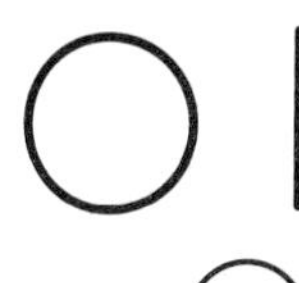

Ⓐ  Ⓒ

2. Is the shape symmetrical?

yes no

 Ⓑ

3. How many equal parts are there?

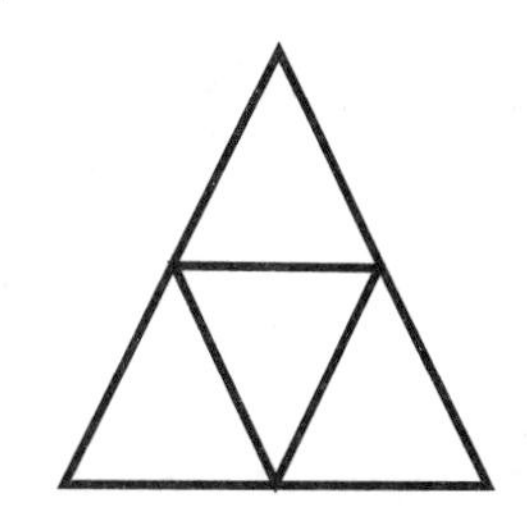

4 5 6

Ⓐ  Ⓒ

4. How much is a penny worth?

10¢ 5¢ 1¢

Ⓐ Ⓑ Ⓒ

5. How much money is there?

9¢ 10¢ 8¢

Ⓐ Ⓑ Ⓒ

6. How much money is there?

15¢ 25¢ 35¢

Ⓐ Ⓑ Ⓒ

Test Practice 3         

Fill in the correct answer bubble.

1. Which one shows 7¢?

Ⓐ Ⓑ Ⓒ

2. Which one has the same sum as 4 + 5?

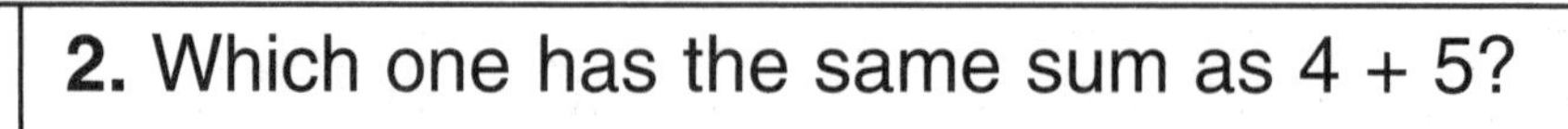

3 + 4	5 + 2	2 + 7
Ⓐ	Ⓑ	Ⓒ

3. Find the difference for 6 – 3.

2	3	4
Ⓐ	Ⓑ	Ⓒ

4. Stu and Lu went to the park at 4:00 and played for 1 hour. What time did they leave?

Ⓐ Ⓑ Ⓒ

5. Band practice began at 2:00 and lasted 3 hours. What time did band practice end?

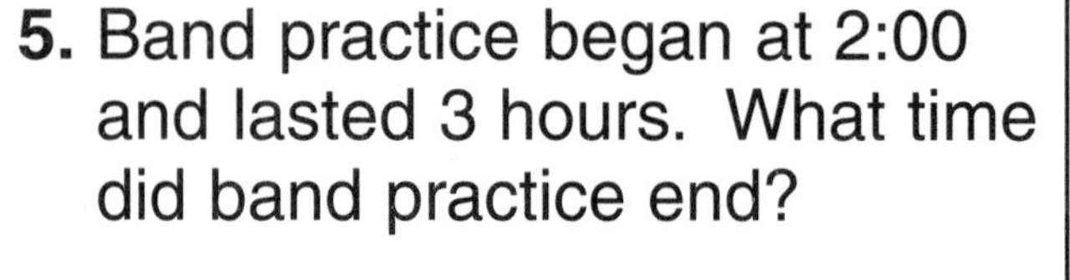

Ⓐ Ⓑ Ⓒ

6. Count the money.

6¢	60¢	16¢
Ⓐ	Ⓑ	Ⓒ

7. Find the subtraction sentence.

–

9¢ – 7¢	8¢ – 1¢	10¢ – 3¢
Ⓐ	Ⓑ	Ⓒ

Test Practice 4

Fill in the correct answer bubble.

1. It is 7:00. What time will it be in 1/2 hour?

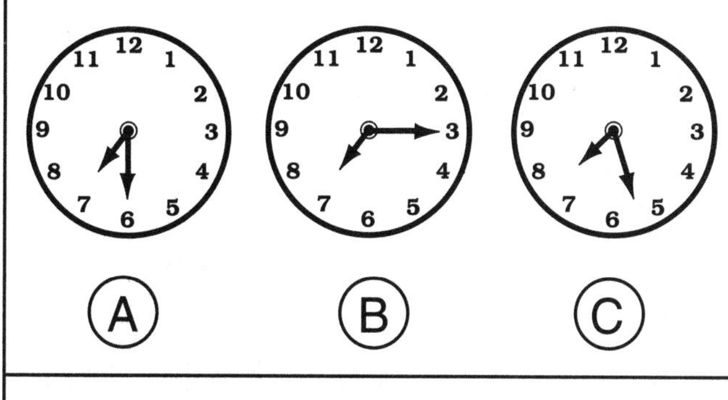

Ⓐ Ⓑ Ⓒ

2. It took the train 1/2 an hour to reach the second station. It arrived at 4:00. What time did it leave the first station?

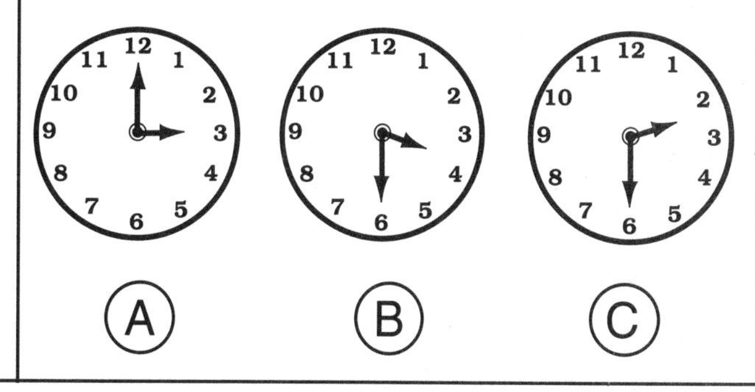

Ⓐ Ⓑ Ⓒ

3. Which set of numbers is in order from smallest to greatest?

5, 8, 13, 25 Ⓐ

25, 5, 8, 13 Ⓑ

13, 8, 25, 5 Ⓒ

4. What is the missing number?

+ 2	5	3	6	0
	7	5	?	2

4 Ⓐ

6 Ⓑ

8 Ⓒ

5. What is the missing number?

4, 6, 8, 10, ____, 14, 16, 18, 20

9 Ⓐ

11 Ⓑ

12 Ⓒ

6. Stacy made 8 cars. Tracy made 5 more than Stacy. How many cars did Tracy make?

3 Ⓐ

13 Ⓑ

12 Ⓒ

Test Practice 5         

Fill in the correct answer bubble.

Fill in the shapes that equal the score.

1. Janie scored 4 points.

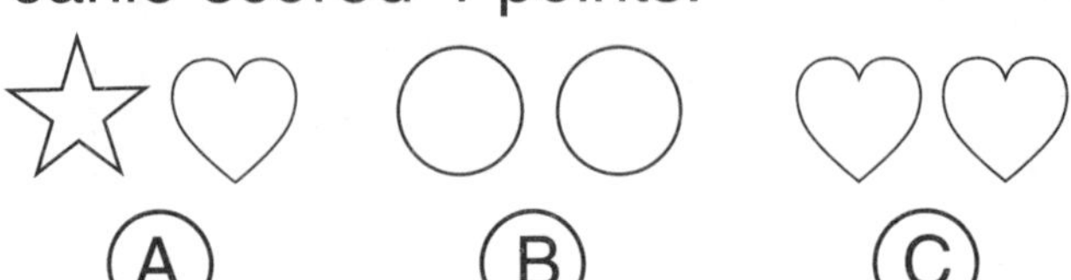

Ⓐ Ⓑ Ⓒ

2. Hank scored 6 points.

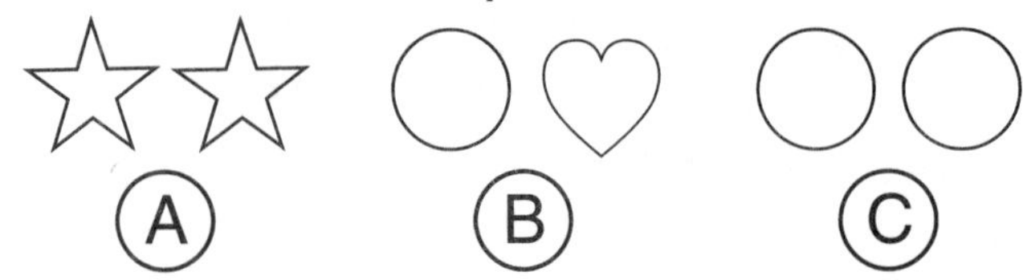

Ⓐ Ⓑ Ⓒ

3. Which addition problem has the same sum as 3 + 5?

4 + 4	6 + 1	7 + 0
Ⓐ	Ⓑ	Ⓒ

4. Which subtraction problem has the same difference as 9 – 5?

4 – 4	7 – 2	6 – 2
Ⓐ	Ⓑ	Ⓒ

5. About how many jellybeans would fill your hand?

1	10	90
Ⓐ	Ⓑ	Ⓒ

6. What is the missing number?

4 + _____ = 10

14	4	6
Ⓐ	Ⓑ	Ⓒ

7. What is the missing number?

8 – _____ = 7

1	0	15
Ⓐ	Ⓑ	Ⓒ

8. Which animal is 3rd?

Ⓐ  

Ⓐ Ⓑ Ⓒ

Test Practice 6

Fill in the correct answer bubble.

1. Fill in the circle under the missing number.

5, 10, 15, _____, 25

16	18	20
Ⓐ	Ⓑ	Ⓒ

2. Fill in the circle under the missing number.

10, 20, _____, 40, 50

25	30	35
Ⓐ	Ⓑ	Ⓒ

3. Fill in the circle under the number that comes before **36**.

35	37	26
Ⓐ	Ⓑ	Ⓒ

4. Fill in the circle under the number that comes after **49**.

48	50	52
Ⓐ	Ⓑ	Ⓒ

5. Fill in the circle under the number that comes in between 57 and 59.

60	56	58
Ⓐ	Ⓑ	Ⓒ

6. Fill in the circle under the largest number.

82	21	37
Ⓐ	Ⓑ	Ⓒ

7. Fill in the circle under the tens and ones that show 52.

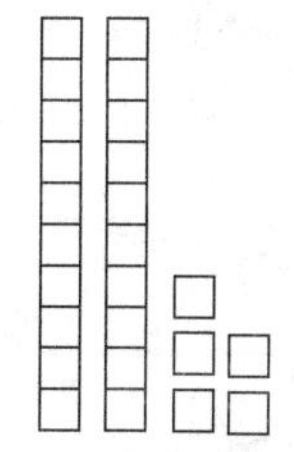

Ⓐ Ⓑ Ⓒ

8. Fill in the circle under the tens and ones that show 20.

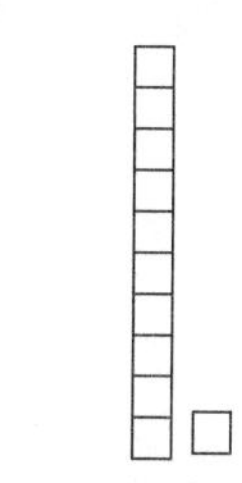

Ⓐ Ⓑ Ⓒ

Answer Sheet

Test Practice 1 (Page 40)	Test Practice 2 (Page 41)	Test Practice 3 (Page 42)
1. (A) (B) (C)	1. (A) (B) (C)	1. (A) (B) (C)
2. (A) (B) (C)	2. (A) (B)	2. (A) (B) (C)
3. (A) (B) (C)	3. (A) (B) (C)	3. (A) (B) (C)
4. (A) (B) (C)	4. (A) (B) (C)	4. (A) (B) (C)
5. (A) (B) (C)	5. (A) (B) (C)	5. (A) (B) (C)
6. (A) (B) (C)	6. (A) (B) (C)	6. (A) (B) (C)
		7. (A) (B) (C)

Test Practice 4 (Page 43)	Test Practice 5 (Page 44)	Test Practice 6 (Page 45)
1. (A) (B) (C)	1. (A) (B) (C)	1. (A) (B) (C)
2. (A) (B) (C)	2. (A) (B) (C)	2. (A) (B) (C)
3. (A) (B) (C)	3. (A) (B) (C)	3. (A) (B) (C)
4. (A) (B) (C)	4. (A) (B) (C)	4. (A) (B) (C)
5. (A) (B) (C)	5. (A) (B) (C)	5. (A) (B) (C)
6. (A) (B) (C)	6. (A) (B) (C)	6. (A) (B) (C)
	7. (A) (B) (C)	7. (A) (B) (C)
	8. (A) (B) (C)	8. (A) (B) (C)

Answer Key

Page 4

1. 3, 1, 2
2. 1, 3, 2
3. 2, 3, 1
4. 3, 1, 2
5. <
6. <
7. <
8. >

Page 5

1. 2
2. 1
3. 3
4. 1
5. 3
6. 6

Page 6

1. 90° F
2. 20° F
3. 40° F
4. 70° F

5. 80°
6. 10°
7. 30°
8. 60°

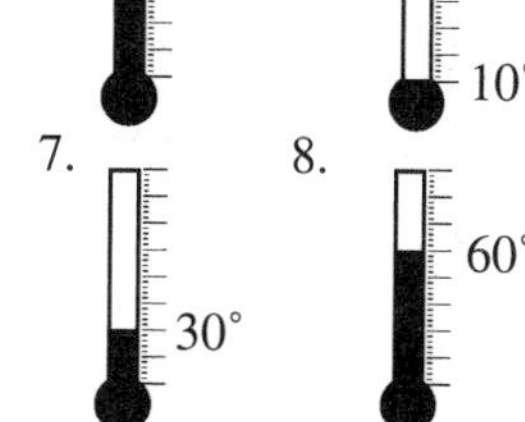

people swimming = 100° F
person gardening = 80° F
person in the rain = 60° F
person shoveling snow = 0° F

Page 7

Across	Down
2. eight	1. zero
4. two	3. ten
5. four	4. three
6. seven	5. five
8. one	6. six
	7. nine

Page 8

A. 9
B. 10
C. 6
D. 6
E. 8
F. 1

1. B
2. C and D
3. F
4.
5.

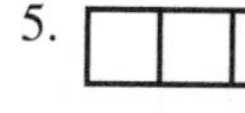

6. or

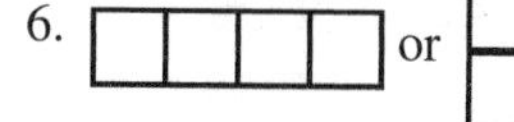

Page 9

1. Color the second and third bear.
2. Color the first and second arrows.
3. Color the first and third stars.
4. Color the second heart.
5. Color the first 3/4 circle shape.
6. square and cube
7. rectangle and prism
8. circle and sphere
9. triangle and pyramid

Page 10

1. heart
2. second triangle
3. down arrow
4. triangle
5. black square
6. black square

Page 11

1. 4
2. 2
3. 3
4. yes
5. yes
6. no
7. no
8. no
9. yes
10. Sample

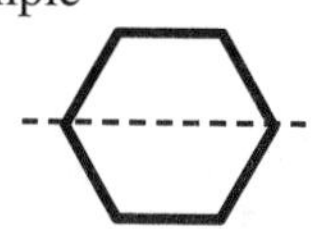

11. Sample

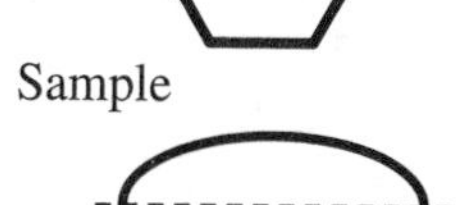

12.

Page 12

1. A
2. C
3. C
4.

5.

Page 13

1. 2¢
2. 4¢
3. 3¢
4. 1¢
5. 5¢
6. 5¢
7. 6¢
8. 7¢
9. 8¢
10. 9¢

Page 14

Table: 10¢, 15¢, 25¢, 30¢, 35¢, 45¢

1. 20¢
2. 25¢
3. 15¢
4. 7 nickels
5. 9 nickels
6. 10 nickels

Page 15

1. 6¢
2. 1¢
3. 8¢
4. 2¢
5. 9¢
6. 4¢
7. 5¢
8. 3¢
9. 7¢

Mystery Message: I can count coins!

Page 16

1. 8¢ – 4¢ = 4¢
2. 5¢ – 5¢ = 0¢
3. 6¢ – 3¢ = 3¢
4. 9¢ – 5¢ = 4¢
5. 4¢ – 2¢ = 2¢
6. 7¢ – 6¢ = 1¢
7. 7¢ – 4¢ = 3¢
8. 6¢ – 6¢ = 0¢

Page 17

Table: 20¢, 30¢, 50¢, 60¢, 80¢

1. 20¢ – 10¢ = 10¢, Herb has 10¢ left.
2. 30¢ + 40¢ = 70¢, Betty has 70¢.
3. 50¢ + 10¢ = 60¢, Dale has 60¢.
4. 80¢ – 70¢ = 10¢, Fay has 10¢ left.
5. 90¢ – 60¢ = 30¢, Sid has 30¢ left.
6. 0¢ + 50¢ = 50¢, May has 50¢.

Page 18

1. 20 pennies = 2 dimes
2. 10 pennies = 1 dime
3. 30 pennies = 3 dimes
4. 60 pennies = 6 dimes
5. 50 pennies = 5 dimes
6. 40 pennies = 4 dimes

Page 19

1. 10:00
2. 12:00
3. 4:00
4. 2:00
5. 1:00
6. 9:00
7.

8.

9.

Page 20

1. walking a mile
2. walking to school
3. washing windows
4. counting $1 in dimes
5. writing a postcard
6. filling the sink
7. 1:00
8. 10:00
9. 6:00
10. 4:00
11. 8:00
12. 12:00

Page 21

1.

3:30

2.

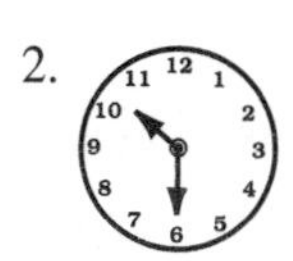

10:30

Answer Key

3. 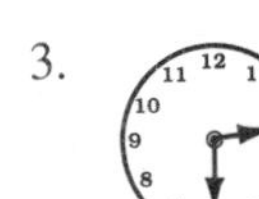
2:30

4. 
4:30

5. 
1:30

6. 
12:30

Page 22

1. 4 mice
2. 5 turtles
3. 6 pigs
4. 5 lions
5. 6 horses
6. 4 frogs
7. Janie
8. Leslie

Page 23

1. >, 10 is greater than 5.
2. >, 9 is greater than 7.
3. <, 7 is less than 10.
4. >, 3 is greater than 2.
5. =, 8 is equal to 8.
6. >, 6 is greater than 1.
7. >, 4 is greater than 3.
8. <, 2 is less than 9.
9. <, 4 is less than 7.
10. =, 7 is equal to 7.

Page 24

1. underline 2, circle 8; 2, 7, 8
2. underline 1, circle 7; 1, 3, 7
3. underline 1, circle 10; 1, 5, 10
4. underline 3, circle 9; 3, 6, 9
5. underline 0, circle 4; 0, 3, 4
6. underline 6, circle 10; 6, 8, 10

Page 25

1. 5, 7, 9, 6
2. 9, 4, 5, 7
3. 10, 9, 7, 8
4. 9, 10, 8, 7
5. 9, 10, 8, 5
6. 7, 9, 10, 8

7. +3 9. +0
8. +1 10. +4

Page 26

1. 8 3. 13
2. 7 4. 16

Color: 2, 4, 6, 8, 10, 12, 14, 16, 18, 20, 22, 24, 26, 28, 30, 32, 34, 36, 38, 40, 42, 44, 46, 48, 50

Page 27

1. 3 3. 1
2. 2 and 2 4. 1 and 1

Page 28

1. 4, 2, 1, 3
2. 2, 3, 5, 0
3. 2, 0, 3, 1
4. 8, 3, 4, 7
5. 1, 7, 2, 5
6. 8, 6, 10, 3

7. –5 9. –2
8. –1 10. –3

Page 29

5—5 + 0; 1 + 4; 0 + 5
7—3 + 4; 7 + 0; 6 + 1
8—5 + 3; 1 + 7; 0 + 8
4—10 – 6; 7 – 3; 4 – 0
6—6 – 0; 10 – 4
2—2 – 0; 4 – 2; 9 – 7

Page 30

1. 7 pennies
2. 1 goldfish
3. 20 marbles
4. 4 teeth
5. 3 minutes
6. 1 hour

Page 31

1. 4 7. 3
2. 7 8. 9
3. 0 9. 5
4. 4 10. 5
5. 2 11. 3
6. 6 12. 5

Page 32

A. 1st (circle)
B. 2nd (heart)
C. 3rd (hat)
D. 4th (glasses)
E. 5th (star)

1. cat
2. bird
3. cow

Page 33

1. Davey
2. the 4th apartment
3. 2
4. 1st
5. 2nd
6. 5 children
7. 7th heart from the left
8. 2nd star from the left
9. 10th arrow from the left

Page 34

Chart: 2, 5, 7
11, 14, 16, 20
23, 27, 28, 29
31, 33, 35, 38, 39
42, 43, 45, 47, 50

1. 32, 34, 23, 43
2. 15, 17, 6, 26
3. 36, 38, 27, 47
4. 18, 20, 9, 29

Page 35

Chart: 52, 54, 56, 58
63, 64, 67, 70
72, 73, 75, 77, 78, 80
81, 84, 85, 88
91, 93, 96, 98, 100

1. 60, 70, 80, 90, 100
2. 51, 52, 53, 54, 55, 56, 57, 58, 59
3. 60, 61, 62, 63, 64, 65, 66, 67, 68, 69
4. 55, 66, 77, 88, 99, 100
5. 53, 63, 73, 83, 93
6. 57, 67, 70, 71, 72, 73, 74, 75, 76, 77, 78, 79, 87, 97
7. any number greater than 86
8. any number less than 74
9. any number greater than 51
10. 91
11. any number less than 60
12. any number less than 65

Page 36

Riddle 1: 40
Riddle 2: 22
Riddle 3: 55
Riddle 4: 20

Page 37

1. 1 ten 5 ones
2. 1 ten 1 one
3. 1 ten 4 ones
4. 1 ten 0 ones
5. 1 ten 9 ones
6. 0 tens 9 ones

Page 38

1. 4 tens 3 ones; 43
2. 2 tens 9 ones; 29
3. 1 ten 6 ones; 16
4. 6 tens 1 one; 61
5. 3 tens 4 ones; 34
6. 5 tens 5 ones; 55
7. 2 tens and 3 ones
8. 3 tens and 8 ones

Page 39

1. 7 4. 12 7. 30
2. 32 5. 10 8. 3
3. 18 6. 31

Page 40

1. B 3. B 5. A
2. A 4. C 6. C

Page 41

1. B 3. A 5. C
2. A 4. C 6. B

Page 42

1. C 4. C 7. A
2. C 5. B
3. B 6. B

Page 43

1. A 3. A 5. C
2. B 4. C 6. B

Page 44

1. C 4. C 7. A
2. C 5. B 8. C
3. A 6. C

Page 45

1. C 4. B 7. B
2. B 5. C 8. C
3. A 6. A